SpringerWienNewYork

Studies in Space Policy
Volume 7

Edited by the European Space Policy Institute
Director: Kai-Uwe Schrogl

Ram S. Jakhu, Tommaso Sgobba,
Paul Stephen Dempsey (eds.)

The Need for an Integrated Regulatory Regime for Aviation and Space

ICAO for Space?

SpringerWienNewYork

Ram S. Jakhu
Paul Stephen Dempsey
McGill University, Montreal, QC, Canada

Tommaso Sgobba
ESA, Noordwijk, The Netherlands

SpringerWienNewYork is a part of
Springer Science + Business Media
springer.at

Typesetting: Thomson Press (India) Ltd., Chennai
Printing: Strauss GmbH, 69509 Mörlenbach, Germany

Cover: Rocketplane Global Inc.

Printed on acid-free and chlorine-free bleached paper

SPIN: 80036708

With 9 Figures and 11 Tables

Library of Congress Control Number: 2011931323

ISSN 1868-5307
ISBN 978-3-7091-0717-1 SpringerWienNewYork

Table of contents

Foreword
Assad Kotaite ix

Executive summary xi
Acknowledgements xv
List of acronyms xvii
List of figures and tables xix

Introduction xxi

CHAPTER 1 Background

1.1 Need for international safety regulations for commercial space activities 3

1.1.1 Introduction 3
1.1.2 Widening access to space and its economic significance 4
1.1.3 Safety Risk of Space Missions 12
1.1.4 Challenges for regulatory regimes and bodies 14

CHAPTER 2 Legal and regulatory regimes

2.1 Current space regulations and standards 21

2.1.1 Legal and regulatory framework 21
2.1.2 Existing International Space Safety Standards 33

2.2 Existing international civil regulatory frameworks, other activities or environments 39

2.2.1 International Civil Aviation Organization (ICAO) 40
2.2.2 International Telecommunication Union (ITU) 43

2.2.3 International Maritime Organization (IMO) 45
2.2.4 Other sources of international law 46

2.3 Transition from air law and space law to aerospace law 49

2.3.1 Introduction 49
2.3.2 Space law conventions 50
2.3.3 Boundary between airspace and outer space 53
2.3.4 Need for a unified legal regime 61

CHAPTER 3 Safety issues

3.1 Safety issues 71

3.1.1 Launch site processing and ground safety 71
3.1.2 Flight hardware, ground support equipment, and COTS 72

3.2 Launch safety 74

3.3 Suborbital safety 79

3.4 Orbital safety issues 82

3.4.1 Orbital debris 82
3.4.2 Collision risk with orbital debris 84
3.4.3 Collision risk to human spaceflight 85
3.4.4 Orbital debris ground risk 87

3.5 Returning vehicles risk 91

3.5.1 Risk to people on the ground 91
3.5.2 Risk to people in aircraft 93

3.6 Saving lives in space missions 96

3.6.1 Extending international search and rescue 96
3.6.2 Ascent emergencies................................ 96
3.6.3 Orbital safety and rescue................................ 97
3.6.4 Ionizing radiation risk for human spaceflight 99

CHAPTER 4 Need for international space safety regulations

4.1 Need for international regulation of STM, space tourism & space debris 103

4.1.1 Commonality or un-commonality of ground standards 103
4.1.2 Comprehensive regulatory approach to space traffic management .. 104
4.1.3 International regulation of aerospace vehicles for space tourism 111
4.1.4 International regulation of orbital debris.................... 113

CHAPTER 5 Proposal for a new regulatory regime

5.1 ICAO for near-space safety? 119

5.1.1 Background 119
5.1.2 Management of space-bound traffic through international airspace... 120
5.1.3 Integration of aviation and space infrastructure................ 122
5.1.4 Integrated terrestrial and space weather forecasts.............. 123
5.1.5 ICAO for an improved international space safety culture......... 124

5.2 Proposal for a new regulatory regime 126

5.2.1 Policy principle................................ 126
5.2.2 Regulatory model 128
5.2.3 Space safety oversight operating model..................... 128
5.2.4 ICAO for space organization........................... 131
5.2.5 General description of the safety certification process 137
5.2.6 Suggested ICAO for space regulatory implementation 138

Appendix A: Relevant excerpts of the ITU constitution and convention 141

Appendix B: Relevant excerpts of the convention on international civil aviation (Signed at Chicago, on 7 December 1944) – Chicago convention 156

Appendix C: Model code of conduct for space-faring nations 178

About the editors. 183

DISCLAIMER: The contents of this Study do not represent the views or opinions of the organizations with which the contributors and reviewers are employed, associated or affiliated.

Foreword

I read with great interest the Study "ICAO for Space" and found it most interesting, well documented and well structured. I am providing in this Foreword some historical background and new perspectives regarding civil aviation.

At the invitation of the United States of America, 52 States met in Chicago and signed, on 7 December 1944, the Convention on International Civil Aviation, known as the "Chicago Convention", which is one of the most remarkable international legal documents of the 20th Century. I like to refer to it as the "*Magna Carta*" of global air transport for its breadth and scope, and for its enduring capacity to ensure the safe, secure and orderly development of what is today certainly the most efficient mode of mass transportation ever created.

This Convention has proven extraordinarily resilient for more than six decades, having been amended but twice in a substantive way, in areas which the visionary drafters of the Convention could not have foreseen (Article 3 bis dealt with the use of weapons against civil aircraft while Article 83 bis addressed the impact of globalization and wide spread economic liberalization of the air transport sector, emphasizing the spirit of the preamble to the Chicago Convention).

However, the 96 Articles of the Convention and its Annexes, which contain close to 10,000 Standards and Recommended Practices (SARPs), are much more complex in nature and relatively arduous to decipher and understand for those not involved in their application. Full and universal compliance with SARPs remains the first condition for maintaining and enhancing the safety of international civil aviation.

Safety, which is the top priority of the Convention, is another common concern we share. Indeed, there is no growth of air transport without safety. In spite of some accidents, air transport is fundamentally safe and remains the safest mode of mass transportation.

However, since this study is entitled "ICAO FOR SPACE", the first time that sub-orbital flights were mentioned in ICAO was at the 35th Session of the ICAO Assembly in 2004 when I said "100 years from now regular passenger flights in sub-orbital space and even outer space could be common place". To date we have no definition where the air space ends and where the outer space commences and, of course, no international treaty was established.

I am of the opinion that there is no need to establish a special international organization for future commercial civil sub-orbital flights, not even for space flights. ICAO is very well structured to meet the necessary requirements for such development in the future by simply extending its mandate to cover this aspect of flights. Although there is no reference in the Chicago Convention to aviation security and environment, nevertheless these two items, together with safety, are top priority in the ICAO Programme and well integrated in ICAO activities. ICAO has developed two Annexes, one for the Environment (Annex 16) and the other for Security (Annex 17). New Annexes could be developed to cover sub-orbital flights and space flights. Should an amendment be needed to cover the sub-orbital and ultimately the outer space civil flights, of course this could be done but it may take a long time for the amendment to enter into force.

ICAO, which was created by the Chicago Convention, remains as relevant a global forum as ever, in promoting the safe and orderly development of international civil aviation. Today we find ourselves in a similar situation with respect to space. With the Chicago Convention we have a model at our disposal. We should not ignore this precious lesson of history by acting expeditiously. We can tackle issues before we are forced to do so.

I commend this Study for its in-depth analysis to all those who are interested in aviation, and wish to express my deepest appreciation to the authors of the study. Their vision will guide the policy of civil flights in space for the years to come.

Assad Kotaite
President Emeritus of the ICAO Council

Executive summary

The rise of the international commercial space sector from low Earth orbits to geosynchronous orbits is transforming the use of space. More actors have increased access for a greater number of activities in space. Yet their proliferation creates a commensurate amount of safety risks – for the general public (on the ground, in the air, and on the surface of the sea), spaceport personnel, space objects, human beings and property in orbit. Environmental accidents pose a threat, as does the ever-increasing amount of space debris and uncontrolled spacecraft re-entry.

There are significant differences between the regimes governing air navigation and space activities. A number of legal issues remain unresolved. Most notably, which regime controls a hybrid vehicle that behaves as an aircraft for one part of its mission and a spacecraft for the other? If a vehicle encounters a problem on the way to space but is still in airspace, to which regimes do those involved look for answers regarding liability? For that matter, where does space actually begin?

This Study addresses the question of whether the extension of the mandate of an existing intergovernmental aviation organization, the International Civil Aviation Organisation (ICAO), is the most appropriate means to initiate and manage regulatory and safety issues for civil and commercial spaceflight up to and including geosynchronous orbits, also considering the growing importance of space-based safety critical services (e.g. for navigation).

To best answer this inquiry, the Study employs the following methodology. First, it describes current regulations and standards bodies that either have developed, or are developing, with regard to space activities, providing an overview of these entities and their activities, be they domestic or international. Next, it assesses the (in)adequacy of the contemporary regime of regulatory protection and promotion of space safety. Further examination is extended to existing international regulatory frameworks in other similar international activities, such as the ITU (International Telecommunication Union) and the IMO (International Maritime Organization) for maritime shipping.

Subsequently, ICAO is analyzed thoroughly and carefully, as it is the entity responsible for promulgating the rules, regulations, procedures and standards that ensure a safe and viable aviation industry. The conflicts between the legal regimes for air and space are identified, including the ongoing functionalist/ spatialist debate and the ambiguity regarding definition of an aircraft and a space object and boundary between air space and outer space. This detailed scrutiny of ICAO includes a discussion of a transition to a new aerospace law, how to extend ICAO's current mandate to include jurisdiction over space activites, and the feasibility of expanding current aviation space traffic management to include suborbital flights.

Finally, in order to understand precisely what a new or extended regime would be regulating, safety issues pertinent to aerospace activities are described in great detail, from launch site processing and ground safety to the launch itself. Ground, orbital, and suborbital risks are addressed, including collision, debris, and traffic management.

The Study led to the following main Findings and Conclusions.

Findings

1. At present, there are no common safety standards and procedures for space operations, thus the public worldwide is not equally protected from the risks posed by launching, over-flying and re-entering space vehicles.
2. Current activities in space are unsustainable in the long term without uniformly implemented debris mitigation measures, well coordinated debris remediation operations, and global space traffic management (STM).
3. The focus of the regulatory regime should be on enhancing the safe and efficient use of space by all actors and the long-term sustainability of Earth orbit without imposing undue restrictions that stifle innovation and commercial development. It should not be so onerous that it undoes benefits for Earth by limiting potential for use.
4. There is no territorial sovereignty or national control in international common spaces such as outer space, the high seas, and international airspace, but only outer space is left without any form of international safety coordination. Furthermore no mutual aid provisions exists for space missions emergencies.
5. It is necessary to traverse airspace to get to outer space. Often this is the international airspace because, many launches occur from locations that are contiguous to the oceans for safety reasons.
6. ICAO already provides ATM, thorough its SARPs, to aircraft in airspace over the high seas (i.e. 72% of the airspace).
7. The prevailing functionality of a vehicle, safety of people on the ground, accumulated knowledge, and best practices in the most closely related fields should drive efforts to classify vehicles.
8. There is a current trend to operate aero-spacecraft from dual-use (airport/spaceport) ground infrastructure.

Conclusions

1. ICAO is a fully experienced and operational legislative and implementing intergovernmental body ideally suited for taking up the issues identified in this Study in relation to aerospace activities.
2. ICAO has in place detailed rules, regulations, guidelines, and operational procedures for aviation that could be gradually extended to space with the necessary modifications.
3. Initially relevant ICAO Annexes should be amended and/or new Annexes should be adopted by ICAO Council in order to address issues such as, *inter alia*, licensing of spaceports, human space flight, space traffic management, safety of personnel and astronauts, and security.
4. Eventually, as the need arises, the Chicago Convention should be appropriately amended to fully establish ICAO's jurisdiction over relevant space activities.
5. It is better to address these issues proactively than retroactively before threats and hazards to public safety become intolerable; now is the appropriate time.
6. A proposed STM regime, to prevent collision between space objects and of space objects with space debris, must be based on a technologically advanced and globally shared space situational awareness system. Such a regime must have its roots in existing international space law, particularly equal rights to space and freedom of use.
7. An international STM organization must be established primarily for the civil and commercial use of outer space and not appended to, or negotiated with, space arms control or disarmament.
8. ICAO's system is sufficiently sophisticated to effectively process these various STM regulatory needs. It is necessary to appropriately classify suborbital (aero-spacecraft) vehicles before they begin flying commercially, though yet difficult to do so because of a lack of standard definitions.

Based upon these Findings and Conclusions, a regulatory model is proposed at the end of this Study, outlining the structure of an ICAO for Space organization and how best it should eventually be established and implemented.

To facilitate extension of ICAO's mandate, the following actions would be helpful:

1. A study of the experience gained by those countries which have already established a national licensing system for commercial space operations should be undertaken.
2. Exploration of methods of linking/merging the ITU information/notification system with an improved UN registration system, with the goal of a unified international notification/information system.

3. Further inquiry into the interests and expectations of private actors and costs and benefits of a global STM system into commercial activities is necessary.
4. A study should be made of the latest trends in technical international organizations regarding the adoption of safety technical regulations/standards, to provide more flexibility than the traditional system of negotiation and ratification.
5. Exploration of policy and regulatory initiatives to achieve and maintain common safety standards and avoid "flags of convenience".

Commencement of these actions would also facilitate timely and smooth introduction of emerging human suborbital and orbital spaceflight international services and eventual implementation of the overall model regulatory regime as suggested by this Study.

Acknowledgements

This book is the result of the cooperative efforts of several experts. It has been prepared under the auspices of the International Association for the Advancement of Space Safety (IAASS) and published with the support of the Institute of Air and Space Law of McGill University, Montreal, Canada. These efforts commenced when the Legal and Regulatory Committee of the IAASS determined that there was the need to explore the possibility of developing international space safety regulations to govern the conduct of commercial space activities. For this purpose, the Committee created the IAASS *ICAO for Space?* Working Group whose work culminated in the production of the first draft of this Study. The members of the Working Group included H. Baccini, Nicholas Bahr, Jerry Haber, Ram S. Jakhu, Paul Kirkpatrick, Kai-Uwe Schrogl, Tommaso Sgobba, J.-P. Trinchero, and Paul Wilde. Subsequently, the draft has been extensively expanded, reviewed, revised, and edited by the three Editors.

The IAASS wishes to express its special appreciation to Nicholas Bahr for initially leading the Working Group and for putting together an earlier draft of the Study, and to: Prof. Kai-Uwe Schrogl (Director of the European Space Policy Institute); Dr. Firooz Allahdadi (U.S. Air Force Safety Centre, Space Safety Division); Dr. Maite Trujillo (European Space Agency); to Dr. Jiefang Huang (Principal Legal Officer of International Civil Aviation Organization-ICAO); Mr. Brian Weeden (Technical Adviser to the Secure World Foundation); Dr. Assad Kotaite (President Emeritus of the ICAO Council) and, Dr. Sanat Kaul (former Representative of India to the Council of ICAO) for reviewing the revised draft and providing useful comments for the improvement of the Study.

Special thanks are also hereby expressed to graduate students of the Institute of Air and Space Law of McGill University, namely: Maria Buzdugan, Diane Howard, Norberto Luongo, Michael Mineiro, Amanda Mowle, Yaw Nyampong, and Susan Trepczynski all of who made various important contributions to the research, proof-reading, and editing of the manuscript for this Study.

We express our deep appreciation to Dr. Assad Kotaite, President Emeritus of the ICAO Council, for thoughtfully writing the foreword for this Study.

Finally, we would like to acknowledge with sincere gratitude the financial support for assistance in research and editing of this Study provided by One Earth Future Foundation, based in Colorado, U.S.A.

The contents of this Study are developed with the intention of initiating international discussion on the subject and do not necessarily reflect the personal views or opinions of the members of the ICAO for Space? Working Group, the editors, researchers and reviewers of this Study. Neither do they represent the official views of any organizations with which they may be associated or affiliated.

Ram S. Jakhu, Co-Editor
Chairman, IAASS Legal and Regulatory Committee,
McGill Institute of Air & Space Law

Tommaso Sgobba, Co-Editor
President, IAASS

Paul Stephen Dempsey, Co-Editor
Director, McGill Institute of Air & Space Law

List of acronyms

A

ADS: Automatic Dependent Surveillance system
ADS-B: Automatic Dependent Surveillance-Broadcast
AIAA: American Institute of Aeronautics and Astronautics
ATM: Air Traffic Management

C

CAIB: Columbia Accident Investigation Board
CEOS: Committee on Earth Observation Satellites
CINA: *Commission Internationale de la Navigation Aérienne*
CoC: Code of Conduct
COPUOS: United Nations Committee on the Peaceful Uses of Outer Space
COTS: Commercial-Off-The-Shelf
CNES: *Centre National d'Études Spatiales* (French Space Agency)
CSG: *Centre Spatial Guyanais* (Guyana Space Centre)

D

DARPA: Defence Advanced Research Projects Agency of the U.S.
DOD: Department of Defence of the U.S.
DSTs: Decision Support Tools

E

EEZ: Exclusive Economic Zone
ELV: Expendable Launch Vehicle
ESA: European Space Agency
EU: European Union
EVA: Extra-Vehicular Activity

F

FAA: Federal Aviation Administration of the U.S.
FAA-AST: Office of Commercial Space Transportation of the U.S. FAA
FSOA: French Space Operations Act of 2008

G

GALILEO: Satellite Navigation System of the EU and ESA
GEO: Geosynchronous (Geostationary) Earth Orbit
GLONASS: Satellite Navigation System of Russia
GNSS: Global Navigation Satellite System

GPS: Global Positioning Systems of the U.S.
GSE: Ground Support Equipment

I

IAA: International Academy of Astronautics
IAASS: International Association for the Advancement of Space Safety
IADC: Inter-Agency Space Debris Coordination Committee
ICAO: International Civil Aviation Organization
ICAN: International Commission for Air Navigation
ISFO: International Space Flight Organization
IMO: International Maritime Organization
ISO: International Organization for Standardization
ISS: International Space Station
ITU: International Telecommunication Union

L

LAAS: Local Area Augmentation System
LEO: Low Earth Orbit

M

MOL: Manned Orbiting Laboratory

N

NAS: National Airspace System
NASA: National Aeronautics and Space Administration of the U.S.
NRC: National Research Council of the U.S.

R

RFI: Request for Information
RCC: Range Commanders Council of the U.S.
RLV: Reusable Launch Vehicle
RORSATs: Radar Reconnaissance Satellites of the Soviet Union
RTS: Radio Thermal Generator

S

SMS: Safety Management System
SAR: Search and Rescue
SARPs: Standards and Recommended Practices adopted by the ICAO Council as Annexes to the Chicago Convention
SATMS: Space and Air Traffic Management System of the U.S. FAA
Space Shuttle: Space Transportation System of the U.S.
SSA: Space Situational Awareness
STM: Space Traffic Management

W

WAAS: Wide Area Augmentation System

List of figures and tables

Figures

Chapter 1 Background

Figure 1.1: Number of Nations and Government Consortia Operating in Space . . . 5

Chapter 3 Safety issues

Figure 3.1: Suborbital rockets altitudes (Credits: © ESA/G. Dechiara) . . . 79
Figure 3.2: Suborbital vehicles configurations (Credits: © Bristol Spaceplanes Ltd and © Canadian Arrow) . . . 80
Figure 3.3: Satellite catalogue growth . . . 82
Figure 3.4: Fallen orbital debris . . . 88

Chapter 5 Proposal for a new regulatory regime

Figure 5.1: Proposed structure and safety roles . . . 130
Figure 5.2: Current ICAO organizational structure . . . 132
Figure 5.3: Proposed ICAO for space organization chart . . . 133
Figure 5.4: Suggested ICAO for space SARPs development process . . . 139

Tables

Chapter 1 Background

Table 1.1: 2010 Worldwide Orbital Launch Activity . . . 9

Chapter 2 Legal and regulatory regimes

Table 2.1: ECSS standards . . . 34
Table 2.2: ISO space safety standards . . . 35
Table 2.3: ISO orbital debris safety standards . . . 36

Chapter 3 Safety issues

Table 3.1: Sample list of ground processing issues 73
Table 3.2: Launch safety risk management 74
Table 3.3: Controlling orbital debris risk 86
Table 3.4: Key re entry safety questions 91
Table 3.5: CAIB recommendations (selected) 92

Chapter 5 Proposal for a new regulatory regime

Table 5.1: Key elements of international space safety regulatory regime 129
Table 5.2: Safety certification process 137

Introduction

The human adventure in space is now more than half a century old. Approximately 6000 lift-offs have taken place and some 500 people have flown into space. Early space programmes were conducted almost exclusively by a few governments for military and civil purposes with little involvement by the private sector. Gradually, commercial uses of space began to develop and now represent the largest share of space activities. The international space community, including the IAASS, has identified the rapid international commercialization of space, particularly in the fields of telecommunication, navigation, Earth observation, and launch services, as an important and positive step for continual global and national economic growth. Recent interest and actions of the private sector in the field of commercial human suborbital spaceflight illustrates the widening range of financial commitments, and business risks the private sector is willing to take in space, especially with SpaceShipOne's aerospace vehicle winning the coveted Ansari X Prize in 2004 by launching up to an altitude of 100 km two suborbital flights carrying weight equivalent to that of three human beings and returning them to Earth two weeks apart. In addition, several governments and business communities around the world are cooperating to help fund the building of civil spaceports. Corporations like United Kingdom-based Virgin Galactic have made firm commitments in pursuing a new suborbital space tourism market and have shown interest in its possible extension to point-to-point international hypersonic travel. At the same time, the Russian and U.S. governments are promoting early steps towards commercial orbital human spaceflight. The Russians flew the first paying orbital space tourist to the International Space Station (ISS) in 2001 and have continued to do so regularly since then. In the meantime, NASA has launched an important initiative to procure commercial transportation services to the ISS.

However, the Shuttle Columbia accident of 2003, a sequence of accidents on ground, (in particular the disaster at the Brazilian Alcantara spaceport in the same year), and various spectacular launch failures have demonstrated the fact that the business of space is still fraught with risks, not only for the crew on board and ground personnel, but also for the public on ground, at sea or travelling by air. Furthermore, the space and ground environments are at risk. Currently, there are millions of objects of various sizes, which pose a direct potential threat to manned and unmanned orbiting space assets, and an indirect threat to space-based terrestrial safety-critical services. There are also important atmospheric effects from chemical rocket propulsion and environmental impacts on ground because of dropping stages and launch failure. For example, in September 2007, the explosion of a Russian Proton M rocket (with toxic fuel weighing about 200 metric tons) contaminated a vast swath of agricultural land in Kazakhstan.

Though the commercial potential of space provides great promise for the global economy, and despite the fact that safety risks are very real and growing, there is no international cooperative effort to protect and enhance global commercial and public interests in space with internationally agreed-upon and enforceable safety risk mitigation standards. Because of this, the IAASS Legal and Regulatory Committee established the *ICAO for Space* Working Group (WG), an independent group of international space safety experts, to study the matter. The mandate of the WG was to prepare this Study and to document and initiate public debate on the need for international space safety standards and a regulatory regime and body for commercial space activities. In addition, the WG deemed it worthwhile to discuss what that body would look like if it is to be built on the model of existing international organizations such as ICAO. This Study is the result of the cooperative efforts of the various members of this Working Group, the reviewers, and others.

The Study discusses the various legal and regulatory instruments, organizations, and standards that currently impact commercial space safety. It notes that the International Standards Organization (ISO) is the only international body that has, thus far, attempted to develop space safety standards for global use. In any case, those standards are unstructured, sparse, generic, and not endorsed by the majority of national regulatory bodies that deal with space activities. Furthermore, they are meant for voluntary use, which defeats the key purposes of achieving an even level of risk mitigation worldwide and preventing unfair competition as a result of the use of substandard safety practices. However, some national and multinational space bodies have developed their own space safety standards. These are the natural reference for any international harmonization effort. In addition, commercial space activities are formally regulated only in a few countries. The U.S., through its Federal Aviation Administration Office of Commercial Space Transportation, is one such example.

Examining international regulatory regimes in analogous industries provides important insights into how such an international space safety regulatory framework might look. For example, the International Telecommunication Union (ITU) regulates radio frequencies and orbital positions. The telecommunications industry found that an international body that can regulate and manage the radio spectrum and orbits was necessary to help the industry grow in a sustainable way. Important search and rescue frequencies are reserved to ensure that they are not negatively impacted by telecoms' spectrum use and growth. The International

Maritime Organization is another example. Again, to support orderly growth of the international maritime industry and the goods and services it provides, it was paramount to establish international safety regulations.

Probably the best analogy is to the International Civil Aviation Organization (ICAO) which was created towards the end of World War II. Countries quickly realized that the global commercial civil aviation industry could not achieve and maintain sustainable growth without an international regulatory framework to ensure that civil aircraft could take off, fly, and land safely anywhere in the world. The adoption and implementation of common international safety standards has made civil aviation one of the most successful and safest modes of transportation. For this reason, and because of the commonality of interests, most notably the sharing of a crowded airspace, this Study focuses particular attention on ICAO as a model, and possibly as a seat (following the example of the Space Transportation Office at the U.S. FAA) for a future international commercial space safety regulatory body. This is why the study is titled "*ICAO for Space*"?

A review of the variety of, and interrelationships between, safety risks that space organizations are facing is important to fully comprehend the challenges envisaged in the effort to create an international space safety regulatory framework. Launch hazards are real and they impact not only those communities contiguous to the launch ranges but also the countries overflown. Orbital and suborbital flights face safety risks from orbital debris as well as the lack of spacecraft traffic management. Spacecraft are exposed to additional risks during their atmospheric re-entry phase.

This Study underlines two points which are discussed in detail and from various perspectives, and which seek to link ICAO to space. The first point finds expression in the thesis that the ICAO organizational model provides a good starting basis for drafting an international civil space regulatory framework. The second is that any international civil space regulatory organization will necessarily share close interfaces with ICAO, in particular for the integrated management of aviation and space traffic through the airspace, and for the safeguarding of future safety-critical aviation systems which will operate in space. Furthermore, a number of hybrid aircraft/spacecraft vehicles may eventually emerge, which would dictate a certain amount of technical and procedural coordination.

This Study is divided into five separate but interconnected parts:

Part A, Background, discusses the general history of spaceflight and the commercialization of space and examines why the international community should consider an organization like ICAO for space.

Part B, Legal and Regulatory Regimes, details the various national and international treaties, organizations, and standards that may impact space safety.

In addition, it analyzes the efficiency of the current regimes (particularly in the aviation and space fields) in place and whether they are suitable to handle the plethora of safety issues related to space activities.

Part C, Safety Issues, details the myriad of safety challenges envisaged during the entire lifecycle of a commercial space vehicle – from launch, through on-orbit operations, to re-entry or disposal. In addition, specific safety issues that need international safety standards are addressed.

Part D, Need for International Space Safety Regulations, makes the case for immediate and expedient international space safety regulations and the need for an appropriate regulatory body to manage them. In particular, space traffic management, orbital debris, and space tourism are discussed. This part also explains how an international set of regulations can help ensure robust space safety programmes for commercial space operations.

Finally, **Part E, Proposal for a New Regulatory Regime**, discusses the governance and operational construct of an international space safety body. In addition, a suggested regulatory operating model, based on expanding the current ICAO mandate to cover "near space", is described and recommended.

CHAPTER 1

BACKGROUND

1.1 Need for international safety regulations for commercial space activities

1.1.1 Introduction

The central enquiry for analysis by this Study is whether ICAO is the most appropriate international organization to regulate the rapidly expanding global space activities, particularly from the space safety perspective. In order to respond to this question, the Study discusses and examines (a) the origin, mandate, structure and functioning of ICAO; (b) the nature, scope, magnitude and trends in space technology and its applications; (c) the regulatory regimes and organizations analogous to those for space activities and (d) the rationale for ICAO's possible involvement in the development of an international regulatory regime governing world-wide space activities, starting with space safety matters.

Towards the end of World War II, the vision and courage which brought into existence the International Civil Aviation Organization (ICAO) changed the face of the aviation industry forever and laid the basis for its astonishing expansion, progress and unequalled safety. Comparing the current state of civil and commercial space developments with the early years of civil aviation (around the 1930s), it is easy to see striking similarities in terms of the difficulty in improving the safety record, lack of traffic management and obstacles to international trade. The early initiatives and steps that eventually led to the creation of the existing international civil aviation regulatory framework bear certain resemblances to the current state of affairs pertaining to the commercial space industry.

A few years after the Wright brothers uncertain "leap" into the air, the Paris Peace Conference of 1919 was convened and the important issue of putting together an international air law code was specifically entrusted to a special Aeronautical Commission of the Conference. Later, an International Air Convention was adopted and it established an "International Commission for Air Navigation" (ICAN). In 1922, a small permanent secretariat was set up in Paris to assist the Commission in its task of monitoring developments in civil aviation and proposing to States measures to keep abreast with those developments.

Due to the tremendous technical and operational advances made in air transportation during World War II, the U.S. initiated studies of post-war civil aviation in 1943. These studies confirmed the widely held belief that civil aviation had to be organized on an international scale else it would not be possible to utilize it as one of the key driving forces for the rapid economic development of the post-war world. In November 1944, the U.S. government hosted 52 countries at the International Civil Aviation Conference in Chicago. The Convention on International Civil Aviation was signed at the end of that Conference on 7 December 1944.

> ICAO was created to set international standards for civil aviation safety

The 96 Articles of the Convention established the basic rules upon which international civil aviation has been conducted since 1944. The Convention established the permanent International Civil Aviation Organization (ICAO), which became a specialized agency of the United Nations in October 1947. It also empowered ICAO to adopt International Standards and Recommended Practices (SARPs) to secure the highest possible degree of international uniformity in regulations and standards, procedures, and organization regarding civil aviation matters. In the ensuing decades, ICAO has become one of the most successful United Nations specialized agencies.

The middle of the 20th century saw the beginning of space programmes that, for more than two decades, would be the exclusive reserve of the competing Cold War Powers (i.e., U.S. and U.S.S.R.). Initially, Europe lagged behind but established a substantial presence later in the century in commercial launch services and space science. The emerging trend for the 21st century is of a global space industry involving the traditional space powers and a multiplicity of governmental and corporate stakeholders worldwide. The space region up to the geosynchronous orbit is becoming more closely associated with commercial and military activities on Earth, while civil government space programmes are focused on distant objectives of interplanetary exploration and science. A number of factors have contributed to usher in this new space age. They include:

- Space technologies and services spreading worldwide;
- Global economic trends and
- Dramatic socio-economic, political, and strategic changes that followed the end of the Cold War.

1.1.2 Widening access to space and its economic significance

Since the dawn of the space age, the number key of players (i.e. military forces, commercial operators, and, in particular, private companies) has exponentially increased and is expected to grow rapidly in the near future as well.

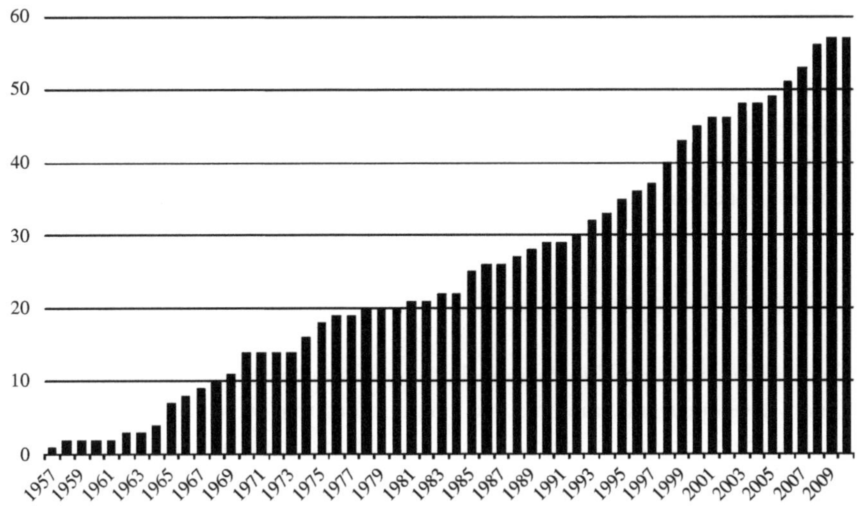

Fig. 1.1. *Number of Nations and Government Consortia Operating in Space.*[1]

The number of nations with unmanned orbital launch capability continues to grow and now includes eleven countries. One may count international operators such as Sea Launch and International Launch Services (ILS) as additional players. A further eighteen countries have acquired suborbital flight capabilities. In February 2009, Iran placed its first satellite in orbit. North Korea maintains a long-range military missile programme that could enable it to quickly develop orbital launch capability. As of December 2010, about fifty countries had accessed space, either with their own indigenously developed launchers or with those developed by other countries, and had assets on-orbit.

> "This revolution is increasingly blurring the traditional distinction between things military and commercial, between things private and governmental, and things domestic and international. We are dealing with a new set of historical conditions, many of them unprecedented. Therefore, we must resist the temptation to apply models or adopt solutions that were more appropriate to the past, or to entirely different historical or economic circumstances without first understanding the implications for the future. . . . This New Space Age or era of "New Space" differs dramatically from the era we have just left in significant ways. First, it is increasingly privately funded and commercial in nature. Second, it will be predominantly international, blurring the once clear lines between what is "ours" and what is "theirs". Keith Calhoun-Senghor, Director Office of Space Commercialization, US Department of commerce to the US House Sub-Committee on Space and Aeronautics, March 1999."

It is interesting to note that in 2003 China became the third country, after the U.S. and Russia, to develop the capability of sending humans into space. Having successfully launched its fist un-manned lunar mission in October 2008, India has announced plans to initiate a human spaceflight programme leading to a first mission in 2014 and a landing on the Moon in 2020. By the end of 2011, the Chinese space station can be expected to be in orbit.

1.1.2.1 Military in Space

From the beginning of the space era, both the United States and the former Soviet Union developed and implemented programmes that were more military than civilian, scientific, or commercial in nature. The first man to orbit the Earth was a Soviet Air Force Major, Yuri Alekseevich Gagarin and he did so on April 12, 1961. The Mercury Redstone rocket which launched the first American, Alan Sheppard, on a suborbital space flight was the product of the U.S. Army Ballistic Missile Agency. The Space Shuttle's technical heritage is deeply rooted in military studies and developments such as the USAF Dyna-Soar of the early 1960s.

The heritage of the international space station can be traced back to military programmes like the Manned Orbiting Laboratory (MOL) and Almaz. In 1963, the USAF initiated the development of the American military space station MOL, which was later cancelled. The Soviet Union responded with the Almaz programme which was a series of military space stations.

Through a tacit agreement between the U.S. and the erstwhile U.S.S.R. which was later codified in certain treaties, overhead reconnaissance by foreign satellites has since the early times been considered an acceptable and legitimate means for confidence building and arms control treaty verification. Military space programmes have also been the precursors of almost all known satellite applications today, ranging from imagery to navigation, telecommunications, and meteorology. For example, commercial use of the U.S. GPS navigation system only started in 1983 following the tragedy in which a Korean civilian airliner was shot down by the Russian air force after accidentally intruding into Russian airspace. Then U.S. President Ronald Regan ordered the USAF to make available degraded GPS signals for civilian use as secondary aid for air navigation. Hitherto, the GPS system had been operated exclusively for U.S. military purposes.

Military use of satellites had become so routine that by 1967 (the year when the Outer Space Treaty came into force and thereby codified the principle of peaceful use of outer space), military satellites were already an integral and irreplaceable part of the defence systems of both the U.S. and the U.S.S.R. In the 1990s, the use of space-based assets became an entrenched and integral component of military planning. The Persian Gulf War of 1991 code named *Operation Desert Storm*, was later described as the first "space war"; it was the first time that the full range of U.S. military space assets were used in active support of actual combat operations on the ground.

States' interest in military use of space has always been, and remains, strong. It will become even stronger as national security becomes increasingly dependent on space-based systems. This strategic interest in the military use of space naturally spurs international debate and the unilateral adoption of policies concerning asset protection and, in turn, "space superiority" as well as "space control". In this respect, a debate is currently raging about limiting the militarization of space to the current balance of applications and forbidding the deployment of space-based weapons. Evaluating and expressing opinions on this debate falls outside the scope of this Study and of the IAASS in general. However, it is important to emphasize that an international civil regulatory framework for space would most probably be accepted by all (military) parties because it might bring significant enhancements to the transparency of civil/commercial operations in space and to the safeguarding of the common space environment and the utilities therein. Military commands have a keen common interest in space traffic management as well as in the control of space debris. They also have an overall interest in transparent communications as a means of preventing military incidents, and they are becoming increasingly anxious about their capability to determine the nature (commercial or military) of satellites on-orbit as the rate of population growth and miniaturization of satellites increases.

1.1.2.2 Rise of international commercial space sector

Already, one sees increased private entrepreneurial activity in space. Since the 1980s, the United States has encouraged "domestic commercial exploration of space capabilities, technology, and systems for national economic benefit".[2] In the U.S., "greatly increased commercial space activity",[3] private sector transportation systems,[4] and the private commercial launch industry are encouraged, facilitated, and promoted.[5]

Early space programmes were conducted almost exclusively by a few governments with little involvement of the private sector, and primarily for reasons of national security and national prestige. It was difficult to justify the economic benefits of space exploration and utilization owing to the immense capital outlay and risk involved. Gradually, however, commercial use of space began to develop as a global industry. In 2009, the global commercial space industry's total revenue (including satellite manufacturing, launch industry, satellite services, and ground equipment manufacturing) reached $262 billion, well in excess of global military space expenditures for that same year.

From an economic standpoint, space commerce has been important for several years. This is due to the traditional satellite telecommunication services and global remote sensing sectors, which have more than tripled in size in the last decade. New commercial fields such as expanded commercial space transportation, enhanced military reliance on space systems, space-based navigation systems, and space tourism are emerging. Navigation systems may become the next pillar of space commerce. Aside from their capability to generate large revenues and profits, civil/commercial space-based systems also play a strategic role as catalysts for further and faster economic and social development on a global scale.

1.1.2.3 Commercial Space Transportation

In the United States, the world's largest market for launch services, space transportation was initially an area in which the government held a monopoly. In the early 1980s, difficulties in meeting an increasing demand, the phasing-out of unmanned expendable launch vehicles (ELVs), and the failure of the shuttle programme to both reduce costs and increase launch frequencies combined to create a substantial commercial market for space transportation in which the European commercial consortium Arianespace took a large share and acted as a role model.

In 1984, the U.S. Commercial Space Launch Act was signed into law. It established licensing and insurance requirements and, for the first time, allowed the use of government property to spur the development of a U.S. based commercial launch industry. The Act also made provision for the U.S. Government to enter into international negotiations to encourage fair competition in launch services. Eventually, as mentioned above, the Shuttle Challenger disaster of January 1986 triggered the complete withdrawal of the U.S. Government from direct provision of commercial launch services. In the meantime, the disintegration of the Soviet Union allowed a number of Russian and Ukrainian companies to enter the launch services market, sometimes in the form of joint ventures with their western counterparts. Recently, China, India, Israel, Japan and South Korea have entered the launch market. (See Table 1.1 on 2010 Worldwide Launch Activities.) Brazil will join this club soon.

In addition to ELV launch services, an interesting new development is taking place within the field of space transportation. Commencing with the suborbital test flights of *SpaceShipOne* which were conducted in 2004, there are at the moment more than 25 different concepts and vehicles under development, eight of which (including the hybrid *Rocketplane XP* based on a modified general aviation

Tab. 1.1: *2010 Worldwide Orbital Launch Activity*[6]

	Commercial launches	Non-commercial launches	Total launches
United States	4	11	15
Russia	13	18	31
Europe	6	0	6
China	0	15	15
Japan	0	2	2
India	0	3	3
Israel	0	1	1
South Korea	0	1	1
Total	23	51	74

Learjet 25) are based on horizontal take-off and landing capabilities. It should be noted that a large majority of suborbital vehicles under development are based on mature technologies and proven operational concepts which are about 40 years old. For example, *SpaceShipOne* and the successor SpaceShipTwo exploited the experience and air-launch flight profile of the USAF X-15 experimental aircraft, which, back in the 1960s, performed tens of flights (sorties), two of which qualified as suborbital flights. (The SpaceShipOne carrier aircraft was named "WhiteKnight" after two legendary X-15 pilots: Robert White and William Knight.) The challenges inherent in the use of such technologies and concepts are those of cost reduction and safety improvement. The former appears to be the main driver and focus of the design efforts, except for public safety-related aspects which are mandated by law.

The "Dragon" is a free-flying, reusable spacecraft being developed by SpaceX under the auspices of NASA's Commercial Orbital Transportation Services Programme. Initiated internally by SpaceX in 2005, the Dragon spacecraft is made up of a pressurized capsule and an unpressurized trunk used for Earth to LEO transport of pressurized cargo, unpressurized cargo, and/or crew members. In December 2008, NASA announced the selection of SpaceX's Falcon 9 launch vehicle and the Dragon spacecraft to resupply the International Space Station (ISS) following the planned retirement of the Space Shuttle in 2011. The first three Dragon demonstration flights should be completed by the end of 2011.

Finally, NASA has launched the demonstration phase of the *Commercial Orbital Transportation Services* programme, the final goal of which is for NASA to contract with one or more private space transport firms to deliver a given amount of cargo to the International Space Station each year. Eventually, this programme would also include transportation and return of crew members. The NASA Commercial Orbital Transportation Services initiative is well in tune with the new U.S. space policy issued in 2010, which clearly enunciates the principle that U.S. government agencies shall use commercial products and services to meet their space programme requirements as first, and not as last, resort. Such policy seems to refer implicitly to programmes in the region of space up to, and including, the geosynchronous orbits.

1.1.2.4 Commercial human spaceflight: emerging new sub-sector of space industry

In the first years of this century, commercial space has started taking the first (but epochal) steps in the ultimate frontier of personal spaceflight.

Orbital space tourism became a reality in April 2001 through the use of Russian government vehicles and related infrastructure. The first space tourist, American businessman Dennis Tito, travelled on a Russian *Soyuz* spacecraft which docked with the International Space Station (ISS). Tito spent eight days in space, six of which were inside the ISS. Tito's successful flight was carried out despite initial objections from NASA and other ISS partner nations over the use of government-owned vehicles. Since then, seven other tourists have visited the ISS. The last one, in 2009, was Guy Laliberte, the Canadian founder of entertainment company Cirque du Soleil.

Manned orbital commercial spaceflight, using privately developed and operated vehicles and ground infrastructure, remains a thing of the future, but important steps have already been taken by two U.S. companies: Bigelow Aerospace and SpaceX. For Bigelow Aerospace, the year 2006 marked a key milestone with the development and launch, on board Russian R36-M launch vehicles, of an unmanned private demonstrator, the inflatable orbiting space station *Genesis I*. *Genesis II* followed in 2007. These two inflatable modules are 30% scaled-down prototypes of a future commercial space station. Once fully-inflated, the commercial space station would be roughly fourteen metres long and seven metres in diameter. These space stations, precursors of the future *Nautilus* commercial space station, were developed on the basis of previous NASA research and patents from the *TransHab* programme, including technologies for space debris shielding. The original plan was to establish a habitable commercial space station for research, manufacturing, entertainment, and other uses by the end of 2011. Currently the "Sundancer", as Bigelow refers to its first full-scale module at present, is scheduled for launch in 2014. The first space tourists would be lifted up to the new commercial space station by 2015 in the CST-100 spacecraft, a crew transport vehicle that is presently being developed by Boeing.

In 2004, SpaceShipOne was the first private "system" to complete two suborbital flights within two weeks, carrying the equivalent of the weight of three adult human beings up to an altitude of about 62.5 miles (100 km) to win the Ansari X Prize. The vehicle was carried by an aircraft up to an altitude of nearly 50,000 feet, released into a glide, and then propelled vertically for eighty seconds by a rocket motor to an altitude of more than 62 miles at apogee, reaching a speed higher than Mach 3. Falling back on its return to Earth, the vehicle re-entered the atmosphere and glided for fifteen to twenty minutes before landing back on the runway of departure. The first commercial suborbital flights are expected to take place sometime in 2011 and will be operated by the U.K. company Virgin Galactic.

Suborbital space tourism operations are expected to begin sometime soon. For instance, a marketing study has estimated that over 15,000 passengers could be flying annually by 2021, representing revenue in excess of US $700 million.[7]

Concurrently, there are plans to establish new commercial spaceports worldwide with new entrants being publicised nearly every year. The most recent announcements in this regard were made by Singapore and Dubai. In perspective, suborbital space tourism may have little to do with space and much with Earth. Vehicles with suborbital flight capabilities have the potential to be used for hypersonic point-to-point international travel. As such, they open up the possibility (sooner than later) for the conduct of "hypersonic-tourism".

Space tourism is but the threshold step in the commercial development of privately financed and built space transportation systems.[8] Once the technology has proven safe for the occasional wealthy tourist eager to float weightlessly and gaze down upon mother Earth, it is likely that entrepreneurs will take the next logical step and employ aerospace vehicles as suborbital transportation vehicles, sharply reducing transit times between the world's major cities.[9] There will be a high-end market for space transportation similar to the high-end business and luxury market for the supersonic Concorde flown by British Airways and Air France. As the move from propeller-driven aircraft to jet-engine powered aircraft revolutionized global transportation, aerospace technology will revolutionize the transportation of persons and cargo around the planet. The suborbital Earth-to-Earth transportation market likely presents a promising long-term commercial opportunity for aerospace vehicles.[10] Over time, it is likely that commercial point-to-point space transportation will eclipse space tourism in commercial importance.

With the promulgation of the Commercial Space Launch Amendments Act of 2004,[11] the U.S. Congress established a strong policy in favour of promoting commercial launches, launch sites, and commercial human spaceflight with necessary regulatory oversight.[12] Other nations are expected to follow the lead of the U.S., both in terms of development of space technology and its applications, as well as their respective national regulatory regimes.

1.1.2.5 Satellite navigation systems

The first global positioning satellite systems (GPS in the U.S. and GLONASS in the Soviet Union) were initially developed for military use. Currently, efforts are far advanced for the development of the first international civilian-controlled satellite navigation system known as Galileo, a joint EU/ESA project. The Galileo technology demonstrator satellite was launched in 2006 and the complete system is expected to become operational by 2014 with a constellation of thirty satellites.[13] In addition to commercial service, Galileo will offer safety of life service as well as search and rescue service. Furthermore, it will make available an encrypted, jam-resistant, public regulated service reserved for public authorities that are responsible for civil protection, national security, and law enforcement. The Galileo system will therefore make available new services for civil and commercial use that were previously not feasible due to military restrictions both in terms of precision of signals and assured continuity of access. China's Beidou ("Big Dipper") geostationary satellite navigation and positioning system is expected to become operational by 2013. Japan and India are also developing their respective global positioning satellite systems. A civilian-controlled satellite navigation system is a strategic means of enhancing the safety of, and expanding the horizons of, commercial aviation. Studies suggest that today's radar-based air navigation system will be strained beyond capacity when demand increases dramatically in the coming years. The implementation of a proposed modern satellite-based air navigation system known as Automatic Dependent Surveillance-Broadcast (ADS-B) will facilitate optimal use of the airspace as it will enable aircraft to broadcast and receive real-time GPS location data directly between themselves and air traffic controllers on the ground and, as such, enable them to fly closer to each other.

1.1.3 Safety Risk of Space Missions

Safety risk in space missions refers to the safety of the general public (on the ground, in the air, and on the surface of the sea), launch range personnel, and human beings orbital and suborbital flights. Space safety also encompasses the safeguarding of valuable assets such as ground facilities (e.g., launch pads), strategic and costly space systems on orbit (e.g., space stations, telecommunications satellites), as well as the safeguarding of the space and the terrestrial environment.

With the exception of the 1986 Shuttle Challenger accident in which the spacecraft exploded during the early ascent phase, the majority of accidents reported under various space programmes and involving human casualties occurred either on the ground or during re-entry. Since the beginning of the era of human spaceflight, 22 astronauts and cosmonauts have lost their lives and this figure constitutes about 4% of the total number of people who have travelled to space. According to official data, to date nearly 200 people have been killed by rocket explosions that occurred during ground processing, launch preparations, and launch operations. Out of the 200 casualties, 35 were counted at the beginning of the 21st century. In the last 10 years, the launch range safety officer, solely for the purpose of preventing risk to the public, terminated at least six launches. It should be noted that this ground safety record is not uniform worldwide. Europe and the U.S. (except for some industrial accidents) have maintained spotless ground safety records. In May 1988, the explosion of a rocket fuel plant at Henderson, Nevada claimed two lives, injuring 372 people and causing property damage estimated at over $100 million over a large portion of the Las Vegas metropolitan area. In July 2007, an explosion at Scaled Composites Inc., a leading player in the private spaceflight industry, killed three people and seriously injured three others.

Environmental accidents such as failures leading to dispersal of radioactive material have also occurred. As of December 2010, there had been ten such cases, including the plutonium payload on board the Apollo 13 lunar module which was jettisoned at re-entry, and which ended up in the Pacific Ocean close to the coast of New Zealand; and the 68 pounds of uranium-235 from the Russian Cosmos 954 which were spread over Canada's Northwest Territories in 1978. The most recent accident of this kind occurred in 1996, when the Russian MARS96 spacecraft disintegrated over Chile, releasing its plutonium payload which has never since been found.

With about 200 "dead" spacecraft abandoned in valuable geosynchronous orbit, orbital debris and uncontrolled spacecraft re-entry pose significant risks. Currently, there are more than 800 operating satellites and 21,000 objects approximately ten centimetres or larger in orbit, not to mention over 300,000 bits of debris with a diameter larger than one centimetre, and many millions more too small to track. Such debris includes pieces of metal, blobs of liquid metal coolant that leaked from discarded space reactors, debris resulting from satellite explosions, and lens covers and other hardware discarded during nominal satellite operations. Some of this material will remain in Earth orbit for hundreds or thousands of years and constitute a potential catastrophic hazard for operational spacecraft because of the high relative velocities at impact.

Debris hazards large enough to be above the tracking threshold are partially controlled by evasive spacecraft manoeuvres. Shielding is also used on manned spacecraft to provide partial protection to those individuals in the habitable modules. Debris impacts on the Shuttle are counted on every mission and samples of residual materials are routinely recovered from their thermal protection systems for examination. The second largest hit ever experienced by the Shuttle was the perforation of a thermal radiator which occurred during the STS-115 mission in September 2006. It did not cause any major problem to the Orbiter, but could have instantly killed an astronaut performing extra-vehicular activities (EVA) at the time if he had been the target.

> Currently, there is no international body that oversees space safety as ICAO does for aviation.

The space industry is rapidly expanding worldwide and with this expansion, safety risks are also increasing at an exponential rate. Some of the reasons for the increased safety risks are inadequate safety oversight, lack of technical progress in space safety, and weak or non-existent domestic and international rules and regulations. There is a high risk to continuing sustained industry growth if safety improvements are not adequately addressed. In particular, a high profile accident with a space tourist on board could immediately cause a moratorium on all commercial launches that include human passengers as part of the payload.

1.1.4 Challenges for regulatory regimes and bodies

Increasing number of actors in space, expanding access to space, the growing volume of space activities, and recently found synergy between civilian/commercial and military users of space, give rise to new regulatory challenges particularly due to the increased safety risks and near non-existence of appropriate international regulatory frameworks.

1.1.4.1 Aerospace vehicles

More than 25 different concepts and aerospace vehicles are currently under study or development. These are mainly in the U.S., but also in Russia, Canada and Europe. Spaceports dedicated to suborbital flights have been, or are in the process of being, constructed in, *inter alia*, the U.S., Singapore, United Arab Emirates, Malaysia, Scotland and Sweden. Several civil aviation authorities have initiated their own studies on possible regulatory frameworks.

The SpaceShipTwo (SS2) vehicles being built by the American company Scaled Composites will be owned and operated by the U.K. company Virgin Galactic. Similarly, the Canadian Arrow suborbital vehicle built in Canada would be operated by the American company Space Adventures. There are plans to operate such vehicles from third-party countries, and, in due course, to use them for point-to-point international supersonic/hypersonic transportation.

Because of such cross-border relationships, complicated legal and regulatory issues arise. For example, with the SS2 classified as a rocket, the U.K.'s Virgin Galactic had to obtain technical assistance agreements under U.S. International Traffic in Arms Regulations (ITAR) to work and exchange data with the U.S. manufacturer Scaled Composites. This translates into additional costs amounting to hundreds of thousands of dollars and long delays in obtaining the required permissions. Then, there are restrictions on what Scaled Composites can disclose to Virgin Galactic (the eventual owner) about the design characteristics of the vehicle.

Furthermore, any country from which SpaceShipTwo will be operated falls within the scope of launching States in accordance with the UN space treaties. Therefore, all such countries will be "*responsible*" and "*liable*" for the space activities that take place from their territory involving that vehicle.

Because the Outer Space Treaty and Liability Convention so require, Virgin Galactic, a U.K. company, has to comply with its native national launch regulations. As such, Virgin Galactic must conform to all requirements prescribed by the British National Space Centre (BNSC), which is involved in the launch licensing process, as well as with those prescribed by the U.K.'s Civil Aviation Authority. Virgin Galactic also has to comply with local (often non-existent) licensing and safety regulations of each country from which it plans to operate SpaceShipTwo. Most probably, Virgin will also need to negotiate dedicated agreements with each of them for insurance coverage of the country's liability in case of accident. It is clear that if commercial human spaceflight is to become a viable international business, coordination and, indeed, harmonization of the national flight safety certification standards, licensing regimes, and liability coverage will become necessary and unavoidable.

1.1.4.2 Space debris

To reduce the risk that space debris poses, satellites have to be disposed of at the end of their operational life either by de-orbiting (those in low orbits) or by moving them to "graveyard" orbits (i.e., those in the geosynchronous orbits). De-orbiting space hardware means that there is still the possibility of debris surviving re-entry and causing damage and/or casualties on the ground. In this case, it is not so much a question of trading one hazard for another because natural de-orbiting would eventually take place due to orbital decay. De-orbiting any spacecraft, or moving it into a graveyard orbit under its own power, requires the use of fuel stored on board the spacecraft. Because there is no international legal obligation to ensure that spacecraft are disposed of properly at the end of their operational lives, some spacecraft operators will rationally choose to use the fuel remaining on board to continue operations and derive substantial profits from the spacecraft instead of using it to place their spacecraft in a safer disposal orbit.

1.1.4.3 Military and civil/commercial space operations

Nowadays, commercial and military space-based systems are synergetic with systems on Earth and, indeed, essential to human activities. Military and civil/commercial space operations need to be clearly separated and their coexistence (in times of peace) adequately managed. Space has become, just as the sea and the airspace, another realm where it is in the interest of the global community to operate in accordance with clear international standards, rules, and procedures instead of imprecise principles.

Use of international waters and the airspace above these waters has been traditionally shared by military forces and commercial operators. Although each pursues its own aims and missions, they have demonstrated an ability to co-exist and co-operate internationally in times of peace and even in war. Military operators have often made available to the general populace valuable services of general interest such as meteorology and traffic control. Currently, international operations in the region of space up to and including geosynchronous orbits are evolving in the same direction. Military and commercial operators remain as the two key players, with the military operating an embryonic space traffic control system, in particular, with reference to orbital debris collision avoidance. In the near future, an international traffic management regime and system will be imperative not only to avoid navigational hazards like space debris, but also to regulate space bound transportation vehicles that will be routinely using free international and controlled national airspaces of various nations. Logic dictates that such a navigational system ought to be integrated with an international air traffic management regime that has essentially already been developed by ICAO.

The rapid and uncontrolled proliferation of orbital space debris – a major threat to spacecraft and the primary risk for human spaceflight – demonstrates the shortcomings of the current regime of voluntary space safety guidelines and codes of conduct. Orbit and frequency allocations, spacecraft and launch vehicle traffic control, safety, and a number of support services such as space weather forecast and orbital debris monitoring, mitigation, and remediation need to be coordinated transparently and effectively both at the national and international levels.

[1] Source: National Air and Space Intelligence Center, cited in National Security Space Strategy-Unclassified Summary, January 2011, page 2.

[2] Ronald Reagan, "National Space Policy", National Security Decision Directive No. 42 (4 July 1982). "The United States Government will provide a climate conducive to expanded private sector investment and involvement in civil space activities, with due regard to public safety and national security." *Ibid.*

[3] National Aeronautics and Space Administration Authorization Act of 1986 § 202(3), Public Law 99-170, 99 Stat. 1012 (Dec. 5, 1985).

[4] National Aeronautics and Space Administration Authorization Act of 1989 § 101(16)(C), Public Law 100-685, 102 Stat. 4083 (Nov. 17, 1988).

[5] Launch Services Purchases Act of 1990, 42 U.S.C. § 2465b.

[6] U.S. Federal Aviation Administration, Commercial Space Transportation: 2010 Year in Review, 3 (January 2011). Please note that this is only one year (i.e. 2010) data.

[7] http://www.spaceref.com/news/viewpr.html?pid=9436 (last accessed: 03 January 2011).

[8] In 2004, Space Ship One became the first privately designed, financed and developed spacecraft to fly humans into suborbital space. Plans soon were announced for the development of a fleet of space vehicles to take tourists into space. Though the trip is quite expensive, reservations are robust.

[9] One must caution that the technology for long-distance suborbital transportation will have to be more exacting than that required for brief weightlessness for tourist travel.

[10] Though different legal rules may govern the launch of space objects into orbit and beyond, or space tourism, or the movement of State aerospace vehicles such as the Space Shuttle from Earth to space to Earth again, this Sub-Chapter evaluates the narrower question of what legal rules may govern the private commercial transportation of passengers from one State to another via space.

[11] Public Law 108-492.

[12] 49 U.S.C. § 70101(a)(5), (6).

[13] "Galileo and EGNOS to drive future road management systems," 19 April 2010, online: http://www.gsa.europa.eu/go/news/galileo-and-egnos-to-drive-future-road-management-systems (last accessed: 03 January 2011).

CHAPTER 2

LEGAL AND REGULATORY REGIMES

2.1 Current space regulations and standards

2.1.1 Legal and regulatory framework

Several entities have developed, or are involved in the process of developing, safety regulations or guidelines that are appropriate for civil space activities both at the national and international levels. This section provides a brief overview of such entities and of their standardization activities with a view to assessing their (in)adequacy regarding regulatory protection and promotion of space safety.

2.1.1.1 National organizations

Numerous countries are actively engaged in space activities either as operators or regulators. The following section samples a few of such countries and briefly describes their space safety regulatory activities.

2.1.1.1.1 United States Governmental Agencies

National Aeronautics and Space Administration (NASA)

NASA was created in 1958 pursuant to the provisions of the National Aeronautics and Space Act,[1] with a mandate to "plan, direct, and conduct aeronautical and space activities". According to section 203(c) of the Act, "[i]n the performance of its functions the Administration [NASA] is authorized: (1) to make, promulgate, issue, rescind, and amend rules and regulations governing the manner of its operations and the exercise of the powers vested in it by law". Among NASA's functions, the role of ensuring the safety of its missions is considered to be of outmost importance. To fulfil this function, two offices within NASA are in charge of developing and implementing safety standards and procedures; i.e. NASA's Technical Standards Programme Office and NASA's Office of Safety and Mission Assurance (OSMA).

(a) NASA's Technical Standards Programme[2] is sponsored by the Office of the Chief Engineer of NASA. Its purpose is to enhance NASA's engineering capabilities by providing the technical standards required by the Agency. Such standards are either NASA-developed technical standards, or non-Government standards adopted by NASA. Some of the NASA standards are mandatory (i.e., they are legally binding). These include the Guidelines and Assessment Procedures for Limiting Orbital Debris,[3] according to which the development of all NASA flight projects must either provide for or incorporate a debris assessment plan.[4] A complete list of the standards cited as mandatory is provided online.[5]

In addition to its own standards, NASA implements the International Standards Organization's (ISO) set of international quality standards (ISO 9000), and is committed to replace NASA's standards with industry standards from the U.S. and other international sources whenever possible in order to improve interoperability and reduce costs associated with aerospace systems.

(b) NASA's Office of Safety and Mission Assurance (OSMA) is in charge of developing and implementing NASA-wide safety and mission assurance (S&MA) policies and standards, and performs independent assessments of programmes and process verification reviews.[6]

Federal Aviation Administration (FAA)

The FAA's primary responsibility is the safety of civil aviation, but the Agency has been mandated additionally to regulate U.S. commercial space transportation. Within the FAA, the Office of Commercial Space Transportation (AST) has the mandate to "ensure protection of the public, property, and the national security and foreign policy interests of the United States during a commercial launch or re-entry activity and encourage, facilitate, and promote U.S. commercial space transportation".[7] Pursuant to this mandate, AST issues launch licenses for commercial launches of orbital rockets and suborbital rockets. A launch – or re-entry – specific license authorizes the holder to conduct one or more launches or re-entries having the same operational parameters of one type of launch or re-entry vehicle operating at one launch or re-entry site. The license identifies by name or mission each activity authorized thereunder.[8] Each licensee is responsible for assuring public safety. A safety review is also conducted by AST to determine whether the applicant can safely conduct the proposed launch operation. It is important to note that, according to the Commercial Space Launch Amendments Act of 2004 (CSLAA), AST is also authorized to issue experimental permits (in lieu of launch licenses) for the launch of, or re-entry of, reusable suborbital rockets for the purpose of research and development to test new design concepts, new equipment, or new operating techniques.[9]

AST also licenses the operations of non-federal launch sites, or "spaceports". On 25 August 2006, the FAA and the U.S. Air Force (USAF) issued new common federal launch safety standards aimed at creating a consistent and integrated system of space launch rules.[10] This approach harmonizes launch procedures that identify potential problems at an early stage and implements a formal system of safety checks and balances.

U.S. Department of Defence (DOD)

The DOD maintains a Defence Standardization Programme, "a comprehensive, integrated standardization Programme linking DOD acquisition, operational, sustainment, and related military and civil communities. It is universally recognized for advancing DOD's Joint Vision 2010 and acquisition goals".[11] Most DOD-issued standards are mandatory. In some instances, standards issued by non-governmental organizations have been adopted by DOD as voluntary standards.

2.1.1.1.2 Non-governmental organizations in the United States

Centre for Space Standards and Innovation (CSSI)

The CSSI is a non-governmental organization whose main purpose is to foster mutually acceptable astrodynamic standards and to "act as a central resource for open, readily available, industry-wide standards to encourage interoperability".[12] The CSSI defines astrodynamic standards as "processes and practices that, by consensus of the world or national industrial and academic communities, best promote interoperability, information exchange, and commerce".[13] Currently, CSSI is developing standards for space systems, re-entry safety control for unmanned spacecraft, and launch vehicle orbital stages. These standards are voluntary but serve as a useful term of reference for the state-of-the art in the field and also, by providing easy access to industry-wide standards, they encourage interoperability.

American Institute of Aeronautics and Astronautics (AIAA)

The AIAA serves the aerospace sector at the national and international standards level by facilitating consensus among government, industry, and academia. The AIAA conducts standards activities through its Committees on Standards[14] and provides an open forum for the discussion of standardization issues of importance to the aerospace profession. The AIAA has published a series of industry (non-governmental) standards that deal with space systems safety (see, e.g., the Standard for Commercial Launch Safety (ANSI/AIAA S-061-1998)).

2.1.1.1.3 Canada

According to the provisions of Canada's Aeronautics Act,[15] the federal government's Department of Transport, also known as Transport Canada, is responsible for regulating rocket launch activities in Canada. According to section 602.43 of the Canadian Aviation Regulations (CAR), "no person shall launch a rocket, other than a model rocket or a rocket of a type used in a fireworks display, except in accordance with an authorization issued by the Minister pursuant to section 602.44". Under section 602.44, "the Minister may issue an authorization (. . .) where the (. . .) launch of the rocket is in the public interest and is not likely to affect aviation safety". The application for such authorization requires that all "rocket activities and participants must comply with the applicable safety codes, launch standards and procedures as developed, published and maintained by a rocketry association accepted by the Minister including all applicable Canadian Aviation Regulations (602.43) and standards as well as other federal, provincial and municipal laws".[16] However, a Type 2A (short form) Risk Assessment will be conducted to determine if an authorization needs to be issued.[17] Regional offices of Transport Canada dealing with general aviation matters are also responsible for providing regulatory oversight of high power rocket launches. Regional staff members are authorized to review applications to ensure that the location and launch activities will be safe and consistent with regulatory requirements. Upon a satisfactory review, Transport Canada issues a launch authorization. Suborbital/orbital rocket launches also require a launch authorization pursuant to CAR 602.44.

2.1.1.1.4 France

With respect to all launches performed from the *Centre Spatial Guyanais* (CSG), located in Kourou, French Guiana, France and the European Space Agency (ESA) are always "launching States" as defined in the 1967 Outer Space Treaty and the 1972 Liability Convention. (ESA has declared its acceptance of rights and obligations under the 1972 Liability Convention, thus the organization is to be treated as a launching State for its space objects.) For SOYUZ launches from

CSG, which will begin in 2011, Russia will also be considered a launching state. According to the terms of several bilateral agreements executed between the French government, ESA and/or Russia, the French space agency, *Centre National d'Etudes Spatiales* (CNES), is responsible for specifying the flight safety rules that are applicable to any launcher and any spacecraft operated from French territory. In consequence, CNES is also responsible for defining, developing, and operating all the means necessary for the safe performance of space flights in relation to all commercial launches performed from the CSG, irrespective of the type of launcher used (i.e., whether it is an ARIANE, SOYUZ, or VEGA).

In order to perform these responsibilities, CNES has established several entities and charged them with specific tasks aimed at guaranteeing genuine independence toward the launch or spacecraft operator, with the support of CNES technical experts:

- a Central Safety Office in charge of harmonizing and coordinating all the safety activities and specifying top level safety regulation (i.e., the CNES Safety Policy), which specifies how CNES assumes and carries out international responsibilities on behalf of the French government. This office reports to the President of CNES;
- a Safety Service at CSG in Kourou in charge of: specifying CSG safety regulations applicable to each flight operator (based on the CNES Safety Policy); verifying that the safety measures and devices related to launchers and spacecraft are in compliance with the safety regulations; and, operating the flight safety system.

Currently, these bilateral agreements also specify for each launcher (i.e., Ariane, Soyuz, or Vega) the manner in which liability for compensation is to be allocated between each of the launching States after a mishap has occurred (e.g.: France and ESA for ARIANE and VEGA launchers, France and Russia for SOYUZ launchers etc.). Moreover, for purposes of simplifying the applicable regime, harmonizing all the French texts associated with the bilateral agreements, assuring coherency for all launches procured by France with or without any other launching State, and promoting space activities in Europe, the French government enacted a Space Operations Act (FSOA) in 2008.

The FSOA was signed into law on June 3rd 2008 and published the next day in the "*Journal Officiel de la Republique Francaise*" as LOI no 2008-518. Its purpose is to set up a national regime to license and control space operations in accordance with the provisions of the relevant international space treaties. Concurrently, CNES' role is extended with reference to the previous CNES Decree no 24-510,

28 June 1984, to include responsibility to issue Technical Regulations under the FSOA. The licensing and control regime basically applies to non-governmental space activities that meet one of the following criteria:

1) *French territory jurisdiction criteria:* when an operator seeks to launch (and/or return) space objects from or to French territory, no matter the actual nationality of the operator of the space object;
2) *French nationality criteria:*

   – when a French national or company headquartered in France (being an operator or not) intends to launch (and/or return) space objects from foreign territory or from a place that is not under any State's sovereignty (e.g., international waters);
   – when a French operator intends to command a space object during its mission in outer space.

To illustrate, a foreign satellite which is to be launched by Arianespace from CSG will not be required to apply for a license because Arianespace is already licensed and controlled by CNES for such (launch) operation, and the fact that the satellite command phase will be performed by a foreign operator will not matter much. On the contrary, a license in accordance with the FSOA will be required in order to command a Eutelsat satellite during its mission in space (Eutelsat being a French company). This will still be the case whether the satellite is launched from French territory or foreign territory.

The FSOA licensing procedure encompasses two components: the general assessment (moral, financial, professional) under the overall administrative authority of the French Ministry of Research, and the technical assessment entrusted to CNES. The CNES technical assessment will be performed in accordance with Technical Regulations that will be issued and maintained by CNES under the FSOA covering:

1) Technical Regulations for launch operations
2) Technical Regulations for satellite operations (in-orbit command and re-entry)
3) Safety Regulations at the CSG

It should be pointed out that in order for CNES to properly exercise jurisdiction and control over the Guiana Space Centre, the original statutes of CNES will have to be modified by the enactment of a specific decree to include responsibility over CSG within the mandate of the President of CNES, acting in the name of the French Government.

2.1.1.1.5 United Kingdom

According to the U.K.'s Outer Space Act,[18] the Secretary of State is authorized to grant licenses for space activities (including launching, procuring the launch of a space object, or operating a space object) as he/she sees fit. In practice, the licensing and other functions entrusted to the Secretary of State are performed through the U.K. Space Agency that was recently created on 23 March 2010.[19] The Secretary of State may not grant a license unless he/she is satisfied that the activities authorized by the license will, *inter alia*, not jeopardize public health or the safety of persons or property. A license may contain conditions, including those that relate to "permitting inspection by the Secretary of State of the licensee's facilities, and inspection and testing by him of the licensee's equipment".[20] The Secretary of State is empowered to make regulations prescribing the form and contents of applications for licenses and for regulating the applicable procedures and standards. In its implementation of the Outer Space Act, the Agency applies several technical space safety standards that are adopted in the U.K., including those that were earlier promulgated by the Secretary of State.[21]

2.1.1.1.6 Russian Federation[22]

The Russian Space Activity Law of 20 August 1993 is the main federal legislation governing space activity in Russia. This law has been amended eight times with the latest amendment occurring in December 2008. In Russia, space activities are carried out in accordance with the Constitution of the Russian Federation and the President of the Russian Federation has overall responsibility for space activities. Russian law imposes the requirement of licensing and safety certification on all federal space operations, including the production and testing of space-rocket complexes and their component parts, storage, preparation for launch, launching, utilization of space vehicles, as well as the control of space missions.[23]

The Russian Space Agency (*Roscosmos*) implements the Federal Space Programme on the basis of the Space Activity Law, Article 22 of which prescribes that all space activities must comply with the safety requirements established by the applicable laws and regulations. The agency is responsible for the implementation of Russian space policy, regulation, the provision of governmental space services, the management of state property for space activities, and, international cooperation. The Agency is also responsible for mandatory certification of rocket and space technologies within the federal system.

The technical standards that are used in the Russian Space Programme are:

- Russian National General Standards, which have mandatory application within the entire territory of the Russian Federation.
- Standards Related to Russian national economic matters, which are mandatory with respect to *Roscosmos* and therefore apply to the Russian Space Programme.
- Plant/corporation-specific Standards, which are mandatory for specific enterprises, for example, the Rocket and Space Corporation "Energya".

The provisions of a series of bilateral agreements concluded between Russia and Kazakhstan in 1992, 1993, 1994, 1999 and 2004 govern safety issues relating to operations conducted at the Baikonur Kosmodrome. In these agreements, both governments have mutually granted the Baikonur Kosmodrome a special legal status; although physically located within the territory of Kazakhstan, the Kosmodrome is considered to be a federal city of the Russian Federation operating under a special legal regime. As such, Russian space safety laws and regulations apply to space operations that take place at the Baikonur Kosmodrome.

According to Olga Zhdanovich, an expert on Russian space safety regulations, the "Russian space safety standards are part of reliability and quality control programmes and distributed along all phases of a life cycle of space technology starting through the development phase, followed by testing and continued through the operation phase. There are not that many special standards for crew health and safety on board of spacecraft".[24] However, she cautions that "Russian safety standards are different by nature and scope from European and the U.S. safety standards. It has been possible because of the different organization of the Russian space industry and historically different approaches for space safety standards development and implementation that take its roots in the Soviet Union times".[25]

2.1.1.1.7 Ukraine

According to Ukrainian space law, State regulation and management of space activity in Ukraine are to be conducted under basic legal principles, standards, and rules governing space activities.[26] Any space facility engaging in, or intending to engage in, space activity within the territory of Ukraine or outside its territorial borders but under its jurisdiction, is required to obtain a license from the Ukrainian National Space Agency authorizing the conduct of such activity.[27] The regulations governing space activities in the Ukraine include operating standards for space facilities as well as standards and regulatory texts governing procedures for, *inter alia*, licensing of space activities, supervision and monitoring of the safety of space launches and flights, and environmental protection in the course of space activity. The regulations governing space activities are established by competent Ukrainian State authorities and are binding upon all Ukrainian persons and entities engaged in space activity.[28]

2.1.1.2 International cooperation

The regulation of space safety has also been the subject of considerable discussion in numerous international fora. In this section we take a look at the various international regulatory approaches that have been or are being considered and the effectiveness of the outcomes that have been achieved so far.

2.1.1.2.1 Inter-Agency Space Debris Coordination Committee (IADC)

The IADC was established in 1993 as an "an international governmental forum for the worldwide coordination of activities related to the issues of man-made and natural debris in space".[29] It comprises the national space agencies of China, France, Germany, India, Italy, Japan, the Russian Federation, Ukraine, the United Kingdom, the United States, plus the European Space Agency. The IADC functions as a forum that allows for the exchange of information on space debris research activities between member space agencies in order to create and enhance opportunities for cooperation in space debris research and to identify debris mitigation options. The IADC makes decisions based on consensus.

In 2001, COPUOS asked the IADC to develop a set of voluntary international space debris mitigation guidelines. On 15 October 2002, the IADC released its guidelines according to which nations should limit debris released during normal space operations, minimize the potential for on-orbit break-ups, undertake post-mission disposal, and prevent collisions.[30] In addition, the IADC recommended as part of the guidelines that a space debris mitigation plan must be submitted for each space mission. It also encouraged nations to voluntarily report on their debris mitigation efforts. On 5 October 2004, the IADC released an explanatory document titled "Support to the IADC Space Debris Mitigation Guidelines" which detailed the purpose, feasibility, practices, and tailoring guide for each recommendation addressed in the Guidelines.[31] The guidelines are voluntary, meaning that they do not have legally binding effect on those who adopt them. The guidelines were submitted by the IADC to COPUOS for consideration at its 2002 session and were subsequently endorsed by the Scientific and Technical Sub-committee of COPUOS in February 2007 (see discussion below, under COPUOS).

2.1.1.2.2 United Nations Committee on the Peaceful Uses of Outer Space (COPUOS)

COPUOS is the main inter-governmental forum for the development of international legal principles governing outer space activities. It was established in 1959 by the United Nations General Assembly (UNGA) under and by virtue of its Resolution 1472 (XIV), and its main purpose is to review the scope of international cooperation in peaceful uses of outer space.[32] The Committee and its two Sub-committees (the Legal Sub-committee and the Scientific and Technical Sub-committee) meet annually to consider issues raised or reports submitted to them by the UN General Assembly and the Member States. The Committee and its Sub-committees work on the basis of consensus and make recommendations to the General Assembly to be considered for adoption as UN resolutions.

In the first 25 years of the space age, COPUOS was very successful in achieving the highest degree of international cooperation in negotiating and adopting the basic international space law treaties which currently govern space activities. They are the 1967 Outer Space Treaty, the 1968 Rescue and Return Agreement, the 1972 Liability Convention, the 1976 Registration Convention, and the 1979 Moon Agreement, which collectively form the foundation of international space law. It should be noted that COPUOS did not adopt any other international space law treaty after the 1979 Moon Agreement. As discussed *infra*,[33] these treaties and agreements are clearly not sufficient to effectively deal with global space safety and traffic management issues in contemporary times.

> COPUOS is the only inter-governmental body that considers all legal aspects of outer space activities.

In terms of its involvement in the field of safety of space activities, COPUOS established the *Working Group on the Use of Nuclear Power Sources in Outer Space* in 1980. This group consisted of national experts and was initially mandated to examine the practice of using nuclear power sources in outer space. In 1983, this mandate was changed and the Working Group was required to develop technical criteria for the safe use of nuclear power sources in space. After almost a decade of discussions, COPUOS submitted a set of principles relevant to the use of nuclear power sources in outer space to the General Assembly. These principles were endorsed by UNGA in its Resolution 47/68 of 14 December 1992.

In light of emerging nuclear power applications and evolving international recommendations on radiological protection, during the 34th session of the Scientific and Technical Sub-committee of COPUOS held in 1997, a decision was made to revive the Working Group on the Use of Nuclear Power Sources (NPS) in Outer Space with a renewed mandate to identify and study the current international technical standards relevant to the use of nuclear power sources. To date, no revisions to the Principles have been recommended. However, at its 47th session held in February 2010, the Scientific and Technical Sub-committee recommended that the Working Group should continue its inter-sectional work on these issues. In addition, the Committee accepted the Work Plan of the Working Group for the period 2010–2015, the objectives of which are promoting and facilitating the implementation of a Safety Framework for Nuclear Power Source Applications in Outer Space, focusing on safety for the relevant launch, operation, and end-of-service phases of space NPS applications.[34]

Space debris mitigation is another area in which COPUOS is working on developing standards. As noted above, a revised draft of the IADC space debris mitigation guidelines was approved by the Scientific and Technical Sub-committee of COPUOS during its 44th session in 2007.[35] "These guidelines are applicable to mission planning and operation of newly designed spacecraft and orbital stages and, if possible, to existing ones." They are not legally binding under international law and it is also recognized that exceptions to the implementation of individual guidelines or elements may be justified. These revised guidelines recommend limiting the amount of debris released during normal operations, minimizing the potential for break-ups during operational phases, limiting the probability of accidental collision in orbit, and minimizing the potential for post-mission break-ups resulting from stored energy. The guidelines were endorsed by the UN General Assembly in December 2007 and are known as the COPUOS Guidelines on Space Debris.[36]

2.1.1.2.3 Committee on Earth Observation Satellites (CEOS)

CEOS' main purpose is to ensure proper international coordination of Earth observation programmes. It consists of various governmental agencies responsible for civil earth observation satellite programmes, as well as agencies that receive and process earth data remotely acquired from space.[37] Currently, CEOS includes more than 25 agencies from the following countries: Argentina, Australia, Belgium, Brazil, Canada, China, France, Germany, India, Italy, Japan, Nigeria,

Republic of Korea, the Russian Federation, Sweden, Thailand, Ukraine, the United Kingdom, and the United States. In addition, more than 20 agencies have associate status in the committee, meaning that they are allowed to participate in all CEOS activities but their vote is not required in order to reach consensus during CEOS plenary deliberations. The activities of CEOS include the development of technical standards for data product exchange. Within CEOS, the Working Group on Calibration and Validation is specifically in charge of addressing the "need to standardize ways of combining data from different sources to ensure the interoperability required for the effective use of existing and future Earth Observing systems."[38] The 31st Plenary Meeting of the CEOS Working Group on Calibration and Validation was held on 2–5 March 2010 in Washington DC, USA.[39]

2.1.1.2.4 European union: proposed code of conduct for outer space activities

Military commands have a keen interest in space traffic management, as well as in the control of space debris. They have also an overall interest in transparent communication as a way of preventing military incidents. The U.S. military have been regularly seeking areas of cooperation with Russia and China that could play a role in helping assuring that the other party does not have ulterior motives when manoeuvring in space. They have been even pondering the development of joint standards for station keeping so that one military is not alarmed by a sudden manoeuvre of the other's satellite that was intended as station keeping, not as the beginning of an attack. In addition for more than two decades a debate has been raging between the U.S. on one side and Russia and China on the other side on banning space weapons. The initial concern was an altered nuclear balance of forces between the superpowers. The debate is better known as Prevention of Arms Race in Outer Space (PAROS), from the relevant draft treaty proposed jointly by Russia and China. The debate started following U.S. President Ronald Reagan's decision to launch in 1985 the Missile Defence Programme (a.k.a. Star Wars), which encountered major technological difficulties and was later restarted in a diminutive form (a.k.a. Son of Star Wars) by U.S. President G. W. Bush after September 11, 2001.

The Code of Conduct (CoC) for Outer Space Activities proposed by the European Union is an international diplomatic initiative to mediate on the above concerns and needs. If successful, it will establish the basic elements of a regulatory framework for outer space operations based on the implementation, through national laws, of guidelines voluntarily agreed by the Subscribing States. At the time of writing at the end of 2010, the CoC was still in the form of a draft being circulated for public comments. The key points are as follows:

a) The CoC is meant to address both space security (military) and space safety. It should be noted that *space security* refers to threats to space systems which are voluntary (i.e. aggressive nature), while *safety* refers to threats that are non-voluntary in nature (design errors, malfunctions, human errors, etc.). As a matter of fact the original concern and main driver of the CoC remains space security, while the aspect of safety has been necessarily included to recognize that the issues related to space debris and space traffic management are common to both fields.
b) The CoC reiterates the principles of "freedom of access" to space for all countries, and the "inherent right to self-defence".
c) The CoC introduces the principle of international cooperative governance of outer space to prevent all kind of interferences. In particular it requires that Subscribing States will promote the development of guidelines for space operations within the appropriate fora for the purpose of protecting the safety of space operations and long term sustainability of outer space activities. It requires in addition the sharing of Space Situational Awareness (SSA) data, and the notification in case of malfunctioning of orbiting space objects with significant risk (e.g. re-entry into the atmosphere or orbital collision).
d) Finally, the CoC establishes on one side the principle that military support systems in space (e.g. GPS, telecommunications, etc.) do not contrast with the principle of peaceful use of outer space, and on the other side that space should not become weaponized in the wide sense (i.e. no deployment of ground-to-space, space-to-space, and space-to-ground weapons).

2.1.2 Existing International Space Safety Standards

Each national space agency has its own set of space safety standards. In addition, numerous multilateral efforts are being made to develop and promulgate space safety standards. This section provides a brief overview of the space safety related activities of concerned international institutions.

2.1.2.1 European Space Safety Standards

European Cooperation for Space Standardization (ECSS) is an initiative aimed at developing a coherent, single set of user-friendly technical standards for use in all

European space activities. The following ECSS standards (see Table 2.1) are related to space-safety.

Tab. 2.1: *ECSS standards*

ECSS-Q-ST-40C Safety (6 March 2009)	• Defines the safety programme and the technical safety requirements that are implemented in order to comply with the ECSS safety policy as defined in ECSS-Q-00 • Intended to protect flight and ground personnel, the launch vehicle, associated payloads, ground support equipment, the general public, public and private property, and the environment from hazards associated with European space systems • ECSS safety policy is applied by implementing a system safety programme, supported by risk assessment
ECSS-Q-ST-40-02 C Hazard analysis (15 November 2008)	• Details the hazard analysis requirements of ECSS-Q-ST-40C • Defines the principles, process, implementation, and requirements of hazard analysis • Applicable to all European space projects where during any project phase there exists the potential for hazards to personnel or the general public, space flight systems, ground support equipment, facilities, public or private property or the environment
ECSS-M-ST-80C Risk management (31 July 2008)	• Defines extending the requirements of ECSS policy principles and requirements for integrated risk management on a space project • Explains what is needed to implement a project-integrated risk management policy by any project actor, at any level (i.e. customer, first level supplier, or lower level suppliers) • Document contains a summary of the general risk management process, which is subdivided into four basic steps and nine tasks • Implementation can be tailored to project specific conditions • Risk management process requires information exchange among all project domains – including safety & mission assurance, and provides visibility over risks, with a ranking according to their criticality for the project; these risks are monitored and controlled according to the rules defined for the domains to which they belong

The European Committee for Standardization (CEN), founded in 1961 by the national standards bodies of the European Economic Community and European Free Trade Area (EFTA) countries, contributes to the objectives of the European Union and European Economic Area by developing voluntary technical standards which promote free trade, the safety of workers and consumers, interoperability of networks, environmental protection, exploitation of research and development programmes, and public procurement. Some CEN standards are directly associ-

Tab. 2.2: *ISO space safety standards*

ISO 14620-1e Space Systems – Safety Requirements – Part 1: System Safety	• Defines the safety programme and the technical safety requirements that are implemented in order to comply with the safety policy as defined in ISO 14300-2 (Space systems – Programme management – Part 2: Product assurance) • Intended to protect flight and ground personnel, launch vehicle, associated payloads, ground support equipment, general public, public and private property, and the environment from hazards associated with space systems
ISO 14620-2 Space Systems – Safety Requirements – Part 2: Launch Site Operations	• Addresses safety liabilities for countries undertaking space activities or allowing operators to perform space activities on or from their territory under outer space treaties adopted by the United Nations • Defines safety responsibilities for operators involved in commercial or non-commercial space launch activities • Establishes overall safety requirements to be observed on a launch site for pre-launch (integration, test, checking, preparation, etc.) and launch operations of a space object • Provides basic principles to enable any operator to implement its own safety methods, tools, and procedures, to ensure the safety of people and personnel, public and private property, and the environment, in a consistent and uniform manner
ISO 14620-3 Space systems – Safety requirements – Part 3: Flight safety systems	• Addresses flight safety systems
ISO17666e Space systems – Risk management	• Defines, extending the requirements of ISO 14300-1 (Space systems – Programme management – Part 1: Structuring of a programme), the principles and requirements for integrated risk management on a space project • Explains what is needed to implement a project-integrated risk management policy by any project actor, at any level (i.e. customer, first-level supplier, or lower-level suppliers) • Contains a summary of general risk management process, subdivided into four basic steps and nine tasks • Implementation can be tailored to project-specific conditions. • Risk management process requires information exchange among all project domains and provides visibility over risks, with a ranking according to their criticality for the project • Risks are monitored and controlled according to the rules defined for the domains to which they belong
ISO146241 to 7 Space systems – Safety and compatibility of materials – Part 1 to 7	• These standards deal indirectly with system safety. For example part 1 deals with the determination of upward flammability of materials • Part 2 deals with determination of flammability of electrical-wire insulation and accessory • Part 4 deals with the determination of upward flammability of materials in pressurized gaseous oxygen or oxygen-enriched environments

ated with the published ECSS standards and the following standards in particular are related to space-safety:

- EN 14738:2004 – Space product assurance – Hazard analysis (CEN publication date 2004-03-24)
- EN ISO 14620-1:2002 – Space systems – Safety requirements – Part 1: System safety (ISO 14620-1:2002) (CEN publication date 2002-12-01)
- EN ISO 17666:2003 – Space systems – Risk management (ISO 17666:2003) (CEN publication date 2003-09-12)

2.1.2.2 International Organization for Standardization (ISO) Standards[40]

ISO is the only truly worldwide organization that has international space safety standards. All others are national, regional or programme standards between national space agencies.

The International Standardization Organization (ISO) is the world's largest developer of standards. The ISO was established on 27 February 1947 with the purpose of facilitating the international exchange of goods and services through the coordination and unification of industrial standards. Representatives of national standards bodies (e.g. ANSI, DIN, etc.) constitute ISO membership on the basis of one member for each country. Currently, ISO has members from 163 countries. Proposed standards are reviewed at the national level through national industrial advisory

Tab. 2.3: *ISO orbital debris safety standards*

The "Orbital Debris Co-ordination Working Group" ISO TC20/SC14 is developing Orbital Debris Mitigation Standards which include:	
Space Debris Mitigation – Principles and Management (ISO 24113)	• Issued in 2010, it defines technical requirements for the mitigation of orbital debris over the life cycle of the space system • It covers all space systems launched into space, including launch vehicle orbital stages, operating spacecraft, and any released objects
Re-entry Safety Control for Unmanned Spacecraft and Launch Vehicle Upper Stages Safety (ISO 27875)	• Rather generic standard, it provides some principles for spacecraft and orbital stages of launch vehicles in order to assess, reduce and control the ground risks when they re-enter into Earth's atmosphere as the consequences of natural decay, intentional reduction of orbital lifetime, or direct re-entry • Supplements the system safety programme specified by ISO 14620-1 from the perspective of re-entry safety

groups supporting their national standard body. The ISO TC 20 SC14 is the ISO Committee responsible for developing standards related to space. Tables 2.2 and 2.3 list some of the ISO space safety standards. With the possible exception of the set of standards being developed in the field of orbital debris, ISO standards are typically not the result of coordinated space policies and initiatives. Most space agencies are not involved in the development of ISO space standards and are not committed to their use. ISO space standards are sparse and often generic. Furthermore, they are intended for voluntary use which is acceptable as long as they respond to purely industrial technical needs. However, in the case of safety standards, voluntarism ends up defeating the key goals of achieving an even level of risk worldwide, prevents the attainment of fair competition due to use of national substandard safety practices, and does not promote proper management of global space traffic. Finally, because safety matters are regulated nationally by laws which also govern the relevant regulatory authority, an unendorsed, international standardization effort like the ISO space safety-related standards tends to be fruitless, despite the fact that they are cooperatively developed and controlled by national space regulatory bodies.

[1] Public Law No. 85568, 72 Stat. 426 438 (29 July 1958) as amended through Public Law 109-155, 119 Stat. 2895, (Dec. 30, 2005).

[2] http://standards.nasa.gov/ (last accessed: 03 January 2011).

[3] NSS 1740.14, online: http://www.hq.nasa.gov/office/codeq/doctree/174014.htm (last accessed: 03 January 2011).

[4] http://orbitaldebris.jsc.nasa.gov/mitigate/mitigation.html (last accessed: 03 January 2011).

[5] http://nodis3.gsfc.nasa.gov/displayDir.cfm?Internal_ID=N_PR_8705_002B_&page_name=main (last accessed: 03 January 2011).

[6] http://www.hq.nasa.gov/office/codeq/ (last accessed: 03 January 2011).

[7] http://www.faa.gov/about/office_org/headquarters_offices/ast/ (last accessed: 03 January 2011).

[8] http://www.faa.gov/about/office_org/headquarters_offices/ast/licenses_permits/launch_reentry/ (last accessed: 03 January 2011).

[9] http://www.faa.gov/about/office_org/headquarters_offices/ast/licenses_permits/sub_orbital_rockets/ (last accessed: 03 January 2011).

[10] Licensing and Safety Requirements for Launch, Final Rule, Federal Register, vol. 71, no. 165 (25 August 2006) at 50508ff. Online: http://edocket.access.gpo.gov/2006/pdf/06-6743.pdf (last accessed: 03 January 2011).

[11] http://www.dsp.dla.mil/APP_UIL/displayPage.aspx?action=content&accounttype=displayHTML&contentid=51 (last accessed: 03 January 2011).

[12] http://www.centerforspace.com/aboutus/ (last accessed: 03 January 2011).

[13] http://www.centerforspace.com/standards/ (last accessed: 03 January 2011).

[14] http://www.aiaa.org/content.cfm?pageid=313 (last accessed: 03 January 2011).

[15] Aeronautics Act (R.S., 1985, c. A-2).

[16] Application for Authorization to Launch High Power and Advanced High Power Rocket(s), http://www.tc.gc.ca/media/documents/ca-standards/26-0659e_0711-03_e.pdf (last accessed: 03 January 2011).

[17] http://www.tc.gc.ca/eng/civilaviation/publications/tp13905-process-1907.htm (last accessed: 03 January 2011).
[18] Outer Space Act (United Kingdom, 1986), 1986, Chapter 38.
[19] For details of the activities and powers of the U.K. Space Agency, visit: http://www.ukspaceagency.bis.gov.uk/Default.aspx (last accessed: 03 January 2011).
[20] Outer Space Act (United Kingdom, 1986), 1986, Chapter 38, Section 5(2).
[21] For a list of the technical standards used in the U.K., see: http://www.ukspaceagency.bis.gov.uk/assets/pdf/Dtbofstnds2010.pdf (last accessed: 03 January 2011).
[22] For details, see Olga Zhdanovich, "Russian National Space Safety Standards and Related Laws," in Joseph N. Pelton and Ram S. Jakhu (eds.), Space Safety Regulations and Standards, 2010, Elsevier, 51 *et seq.*
[23] Statute No. 104 – Statute on Licensing Space Operations, 2 February 1996.
[24] Olga Zhdanovich, "Russian National Space Safety Standards and Related Laws," in Joseph N. Pelton and Ram S. Jakhu (eds.), Space Safety Regulations and Standards, 2010, Elsevier, 51 *et seq.*
[25] *Ibid.*
[26] For a detailed discussion of regulation of space activities in Ukraine, see Nataliya Malysheva, "Regulation of Space Activities in Ukraine", in Ram S. Jakhu (ed.), National Regulation of Space Activities, (Springer, 2010), 335 *et seq.*
[27] Ordinance of the Supreme Soviet of Ukraine on Space Activity, Law of Ukraine of 15 November 1996 (VVRU, 1997, p. 2), Licensing of Space Activity, Article 10.
[28] Regulation Governing Space Activity, Article 8.
[29] http://www.iadc-online.org/index.cgi (last accessed: 03 January 2011).
[30] Available on line: http://www.iadc-online.org/index.cgi?item=docs_pub (last accessed: 03 January 2011).
[31] http://www.iadc-online.org/index.cgi?item=docs_pub (last accessed: 03 January 2011).
[32] http://www.unoosa.org/oosa/COPUOS/copuos.html (last accessed: 03 January 2011).
[33] See Section 4.2 *infra.*
[34] Safety Framework for Nuclear Power Source Applications in Outer Space, UN Doc. A/AC.105/934 (19 May 2009).
[35] Official Records of the General Assembly, 62nd Session, Supplement No. 20 (A/62/20), paras. 117 and 118 and Annex (2007).
[36] United Nations General Assembly, 62nd session, Agenda item 31, Document A/RES/62/217 (10 January 2008), paragraph 26.
[37] http://www.ceos.org/ (last accessed: 03 January 2011).
[38] http://www.ceos.org/index.php?option=com_content&view=category&layout=blog&id=163&Itemid=246 (last accessed: 03 January 2011).
[39] http://www.ceos.org/images/wgcv/wgcv31/wgcv-31_minutes_v1.0.pdf (last accessed: 03 January 2011).
[40] ISO, online: http://www.iso.org/iso/home.html (last accessed: 03 January 2011).

2.2 Existing international civil regulatory frameworks, other activities or environments

The International Civil Aviation Organization (ICAO), the International Telecommunication Union (ITU), and the International Maritime Organization (IMO) are all examples of well-established international organizations with memberships cutting across the vast majority of States. Each of these organizations has power to set standards within its area of responsibility; while none of them has specific and strict enforcement powers, those standards promulgated by them are widely and consistently followed. The key to widespread international participation in these organizations, as well as the corresponding adherence to the standards the organizations put forth, is likely rooted in the international character of the activities in question. Aviation, telecommunications, and shipping are all, by their very nature, international activities. It has, therefore, long been recognized by the States parties to the various international conventions governing these activities that it is in the interest of all States to cooperate in setting universally applicable international standards to regulate these activities.

The standards set by ICAO are critical to international aviation. Without standards and air traffic rules and regulations, it would be very difficult, likely impossible, to sustain a safe and viable international air transport industry. The same can be said about the importance of the IMO and its standards/regulations to international shipping. While the ITU standards may not have the same sort of front line relevance with respect to human safety, they are nonetheless critical, as the significance of reliable, interference-free international communications to almost any international activity, including space exploration and utilization, cannot be discounted.

As previously mentioned, the key seems to be the recognition of the importance of standards by the international community. Each of these organizations works because the member States believe in the necessity of the standards it puts forth, and the member States individually derive some benefit from those standards. Thus, the most important first step in creating international standards to govern a new realm such as commercial space is to make the international community recognize the necessity of comprehensive, consistent, international standards to the continued growth of the industry. While the need for international standards may be obvious, especially given the undeniable technological similarities between the aviation and space industries, two facts may make it more challenging to persuade the international community regarding the need for, and the benefits of, a system of uniform international space safety standards and regulations. They are, first that commercial space (which although started in 1965 with the launch of the INTELSAT 1 satellite on board a Delta rocket) has still not reached its maturity and, second and perhaps more significantly, the fact that it is not something that the majority of States yet participate in.

2.2.1 International Civil Aviation Organization (ICAO)[41]

The establishment of the International Civil Aviation Organization (ICAO) commenced in 1944 with the adoption of its Constitution, the Chicago Convention,[42] and it ended in 1947. At present, 190 States have ratified the Chicago Convention. ICAO is a specialized body of the United Nations (UN), with a mandate to ensure the safe, efficient, and orderly development of international civil aviation. ICAO is composed of three organs: (a) the Assembly, made up of all contracting States; (b) the Council, composed of 36 States and (c) the Secretariat which consists of five bureaux: the Air Navigation Bureau, the Air Transport Bureau, the Technical Co-operation Bureau, the Legal Affairs and External Relations Bureau, and the Bureau of Administration and Services, as well as other offices. The Council, constituted by the official representatives of thirty-six member States elected every three years, is the permanent governing body of ICAO. It is the body that is entrusted with power to adopt and amend standards and recommended practices (SARPs), which are, for convenience, designated as "Annexes" to the Chicago Convention. It is therefore significant to note that without any formal amendment to the Chicago Convention, the Council can make appropriate changes to the existing ICAO Annexes and thus expand ICAO's jurisdiction over space matters, (for starters those that have an impact on the safety, regularity and efficiency of civil aviation).[43] The Council has several committees including: the Air Transport Committee, Finance Committee, Committee on Joint Support of Air Navigation Services, Technical Cooperation Committee, and Unlawful Interference Committee. It should be noted that for achieving ICAO's goals, all the ICAO bodies work cooperatively, though some of them originated from the provisions in the Chicago Convention and others have been created in accordance with the procedures of the ICAO Council.

The SARPs cover all technical and operational aspects of civil aviation, thereby providing the basis for maintaining uniformity in the regulation and safety of civil aviation. A "standard" is "[a]ny specification for physical characteristics, configuration, material, performance, personnel or procedure, the uniform application of which is recognized as necessary for the safety or regularity of international air navigation and to which Contracting States will conform in accordance with the Convention; in the event of impossibility of compliance, notification to the Council is compulsory under Article 38 of the Convention".[44] A "recommendation" is "any specification for physical characteristics, configuration, material, performance, personnel or procedure, the uniform application of which is recognized as desirable in the interest of safety, regularity or efficiency of international air navigation, and to which Contracting States will endeavour to conform in accordance with the Convention".[45]

The SARPs found in the Annexes to the Chicago Convention contain only essential regulatory material; detailed technical specifications are spelled out in attachments to the Annexes, or in separate procedures and manuals. For example, the very technical and detailed Procedures for Air Navigation Services (PANS), which are often elaborations of basic principles found in the corresponding SARPs, are listed separately. Also separate from the SARPs are the Regional Supplementary Procedures (SUPPS), which, unlike SARPs, do not apply generally but only in designated regions.

Technical SARPs, the primary source of ICAO "regulations", are generally formulated in the Air Navigation Commission (ANC), which has been created pursuant to the provisions of the Chicago Convention. The ANC consists of nineteen individuals who are nominated by member States and appointed for three years by the Council; the ANC members act in their personal expert capacity, not as representatives of the States that nominate them. After the ANC formulates the SARPs, it passes them on to the Council for review and adoption.[46] A two-thirds majority of the membership of the Council must vote in favour of a proposed new or amended SARP in order for it to be adopted, even though after such approval and adoption member States may register their disapproval. States have sixty days for filing differences. If a majority of States does not register its disapproval by the effective date, the SARPs enter into effect. In practice, however, standards are not necessarily binding on all States because any State is allowed to file a difference to any SARP, if they find it impracticable to comply with.

Strictly speaking ICAO generally does not possess enforcement power; States that have not notified ICAO of differences between their national regulations and the contents of the SARPs bear responsibility for implementing SARPs once they become effective. However, in order to promote the safety of aviation, ICAO established the Universal Safety Audit Programme in 1999 and, under it, contracting States agreed to "regular, mandatory, systematic and harmonized safety audits",[47] which are carried out by ICAO. In 2002, ICAO established a similar procedure with respect to security (i.e. the Universal Security Audit

Programme) in order to promote global harmonization of aviation security measures through the auditing of Contracting States on a regular basis on their implementation of ICAO Annex 17 – Security.[48]

It is interesting to note that as early as 2000, Dr. Assad Kotaite, then President of the ICAO Council, asserted the pertinence of ICAO's involvement in space transportation.[49] Moreover, it is not widely known, but the ICAO Council already has embarked upon consideration of some space-related matters. For example, during the 175th session of ICAO Council, a Secretariat Working Paper numbered C-WP/12436 dated 30th March, 2005, was presented by the Secretary General and considered by the Council. The Working Paper went into all the details of the Chicago Convention, Assembly Resolutions, various space Conventions, issues of space objects, issues of sovereignty, and the U.S. Commercial Space Launch Amendment Act, 2006, and suggested that future suborbital flight should be subject to International Air Law and the standards contained in the Annexes to the Chicago Convention. The Working Paper concluded that:

> "[v]ehicles which would effect Earth-to-Earth connections through suborbital space could incorporate the constitutive elements of aircraft and fly as such at least during descending phase while gliding. However, rocket-propelled vehicles could be considered as not falling under the classification of aircraft. At this stage, one State seems to prefer to classify such vehicles as rockets . . . . The Chicago Convention applies to international air navigation but current commercial activities envisage suborbital flights departing from and landing at the same place, which may not entail the crossing of foreign airspaces. Should, however foreign airspace (s) be traversed, and should it be eventually determined that suborbital flights would be subject to international air law, pertinent Annexes to the Chicago Convention would in principle be amenable to their regulation."[50]

It is therefore reasonable to predict that, in due course, ICAO not only should but would expand its jurisdiction to cover space-related matters by slowly starting with those space activities that affect the safety of civil aviation. The Chicago Convention vests in ICAO ample jurisdiction to address these critical issues. Article 37 authorizes the promulgation of SARPs addressing "such other matters concerned with the safety, regularity, and efficiency of air navigation as may from time to time appear appropriate". Under this provision, for example, ICAO has promulgated Annexes addressing environmental issues and aviation security, areas not contemplated when the Chicago Convention was originally drafted in 1944. It is manifestly desirable for the same essential rules of safety and navigation to be applied to all users of common airspace – aircraft and aerospace vehicles and space objects on launch and re-entry, as a first step. The extension of these regulations to the geosynchronous orbit would be a desirable second step.

2.2.2 International Telecommunication Union (ITU)[51]

The International Telecommunication Union (ITU) and its predecessor organizations have been in existence since 1865 when the first International Telegraph Union Convention was adopted. Since that time, the ITU has grown to become a specialized agency of the United Nations with a mandate over the international regulation of all forms of wired and wireless communications. The ITU is governed by the ITU Constitution and Convention, the latest version of which was created following structural reforms that were implemented in 1994. It was subsequently amended in 1998, 2002, 2006 and 2010.

The Constitution and Convention are complemented by Administrative Regulations adopted by the ITU in the form of Radio Regulations and International Telecommunication Regulations. While the Constitution and Convention, as well as the Administrative Regulations, are binding on all ITU member States, the ITU has no specific enforcement powers. Its rules and regulations are, on the whole, respected by ITU member States, the broad majority of which believe that it is in their self-interest to do so. However, there is an Optional Protocol to the Constitution, Convention, and Administrative Regulations that provides for compulsory arbitration of disputes regarding the interpretation of the Constitution, Convention, or Administrative Regulations. The ITU thrives on cooperation between governments and the private sector. Private entities (known as Sector members) are allowed membership within the ITU and they include telecommunication "carriers, equipment manufacturers, funding bodies, research and development organizations and international and regional telecommunication organizations".[52] Currently, the ITU membership consists of 192 States members and more than 700 Sector members and associates.

The highest-level decision-making body of the ITU is the Plenipotentiary Conference, which is held once every four years to determine the activities and direction of the ITU. The last ITU Plenipotentiary Conference took place in the city of Guadalajara (Mexico) from 4 to 22 October, 2010. In the period between Plenipotentiary Conferences, the ITU Council meets regularly, acting as the ITU governing body and ensuring the smooth day-to-day functioning of the Union. The ITU Council, presently made up of 48 member States, is also responsible for broad policy issues, the ITU strategic plan, coordination and day-to-day operations of the ITU, and the budget and finances of the ITU. The Council is also the body with authority to implement the provisions of the ITU Constitution and Convention, the various regulations, and decisions made by conferences or meetings of the ITU and its Sectors.

The substantive work of the ITU is divided between three sectors: the Radiocommunication Sector (ITU-R), the Telecommunication Standardization Sector (ITU-T), and the Telecommunication Development Sector (ITU-D). Each Sector has its own bureau, which coordinates daily activities and implements the Sector's work plan. With the support of numerous study groups, the work of the various Sectors result in the making of ITU Recommendations; each year, a total of over 500 new and revised Recommendations are produced by the three Sectors acting in combination.

The ITU-R "draws up the technical characteristics of terrestrial and space-based wireless services and systems, ... develops operational procedures,"[53] including procedures for the avoidance of harmful interference, and conducts technical studies which help guide regulatory decisions such as the content of the Radio Regulations. The Radio Regulations, which govern the use of the radio frequency spectrum, are created and/or modified by international negotiations that take place at world radio communications conferences, which are held regularly every two to three years. The ITU-R is responsible for managing the radio-frequency spectrum and acts as the registrar for international radio frequency use, including maintaining the Master International Frequency Register.

The ITU-T creates technical and operating standards for international telecommunications by studying technical, operating, and tariff questions and adopting recommendations on them with a view towards standardizing telecommunications on a worldwide basis.[54] International telecommunication standards are created through the ITU's world telecommunication standardization conferences and telecommunication standardization study groups (the majority of whose membership consists of private sector members of the ITU) as well as the Telecommunication Standardization Bureau. The ITU-T focuses on the rapid development of effective international standards that are responsive to current technological developments. Currently, the ITU-T creates about 210 new and revised recommendations each year, which equates to one standard for every working day. Included among the many areas covered by ITU-T recommendations are those dealing with communications security, an increasingly important consideration given the growing importance of global communication systems.

The ITU-D's task is "to help spread equitable, sustainable and affordable access to telecommunications, as a means of stimulating broader social and economic development".[55] At present, about two-thirds of the 192 ITU member States lack access to basic telecommunication services, underlining the importance of the work done by the ITU-D. To facilitate its work, the ITU-D emphasizes public–private partnerships which draw on the strengths of private industry to help meet the needs of developing nations.

2.2.3 International Maritime Organization (IMO)[56]

The International Maritime Organization (IMO) is the UN specialized agency tasked with developing and maintaining legal and regulatory standards for international shipping. Maritime safety is thus the primary focus of the IMO. The IMO was formed in 1948 pursuant to its charter treaty, the IMO Convention,[57] which entered into force in 1958. Pending the entry into force of the IMO Convention, the UN Transportation and Communications Commission had the responsibility and Dr. Assad Kotaite served as its chairman for 2 years. The IMO currently has 169 member States, each of which contributes to the IMO budget; the contributions of each State are based primarily on the tonnage of its merchant fleet.

The IMO is composed of an Assembly, a Council, four main committees, and numerous Sub-committees. The Assembly, which meets regularly once every two years (and may meet in additional extraordinary sessions), has the exclusive power to make recommendations to States regarding maritime safety and pollution prevention. The Council, the executive branch of the IMO, is elected every two years at the meeting of the Assembly. The duties of the Council include coordinating the activities of the various organs of the IMO and commenting on reports and proposals from the various committees and placing them before the Assembly.

The most important committee of the IMO, and also the most important from a safety perspective, is the Maritime Safety Committee (MSC), which is composed of all IMO member States. Assisted in its work by nine sub-committees,[58] the MSC considers "any matter within the scope of the Organization concerned with aids to navigation, construction and equipment of vessels, manning from a safety standpoint, rules for the prevention of collisions, handling of dangerous cargoes, maritime safety procedures and requirements, hydrographic information, log-books and navigational records, marine casualty investigations, salvage and rescue and any other matters directly affecting maritime safety".[59] The MSC then submits its work product, in the form of safety-related recommendations, guidelines, and/or draft conventions, to the Assembly for adoption.

At the time the IMO was established, there were already a number of significant international conventions governing maritime activities, including the safety aspects thereof. The IMO was tasked with the responsibility of ensuring that the existing conventions were kept up to date, and it was also given the power to create new conventions as and when necessary. The IMO currently administers a comprehensive body of international conventions supplemented by numerous regulations covering all aspects of maritime transport. Given the rapid rate of change in technology associated with marine transport, amendments to these conventions are often necessary. In order to streamline and speed up the amendment process, many IMO conventions incorporate a "tacit acceptance" procedure which allows an amendment to enter into force after a set period of time not less than one year, unless, prior to that fixed date, a certain percentage of States' party to the convention register their objection. This eliminates the delay often associated with requiring a percentage of States to positively accept amendments which often prevents amendments from ever taking effect.

While conventions are binding on those States which are parties to them, IMO Codes and Recommendations are not mandatory, though member States are expected to implement them. The IMO itself has no enforcement powers; member States are responsible for enforcing provisions of the various conventions with respect to their own ships and, in some instances (especially in their jurisdiction), may have limited powers with respect to the ships of other States. In some cases, the conventions require that certifications be carried on board a ship, signifying its compliance with the terms of the convention. While not a direct enforcement mechanism, the IMO recently introduced the Voluntary IMO Member Audit Scheme, which allows a member State to request an IMO audit of the effectiveness of its administration and implementation of mandatory IMO instruments.

2.2.4 Other sources of international law

As a discipline, international law draws from two major sources of law: customary law and convention-based or treaty law. Customary international law is created through the practice of States. Initially, customary international law made up much of the body of international law. Today States often enter into agreements between themselves, creating convention-based international law. Many international conventions codify existing principles of customary international law, though, of course, customary international law is generally applicable to all States, regardless of whether a particular State is party to a treaty recognizing a given customary principle.[60] Therefore, while general principles of customary international law may be incorporated into various conventions, the applicability of those principles is not limited to parties to those conventions, or even to the general subject matter of the conventions.

While different subjects are covered by different treaties, there are many overarching principles that are found in multiple treaties; the basis of many of these principles is customary international law. One of the well-recognized principles of customary international law included for example, in many international environmental law conventions, is the prohibition against trans-boundary harm, meaning a State cannot allow its territory to be used in a manner which causes injury to another State.[61] As is the case with many general principles of customary international law, the prohibition against trans-boundary harm is significantly broader than individual conventions, environmentally-based or otherwise, that may incorporate it.

Principles of customary international law can also evolve from principles that were originally included in conventions or Resolutions of the United Nations General Assembly, provided that widespread and virtually uniform adherence to these principles is achieved, and States generally recognize the principles as legally binding.[62] Arguably, many of the existing conventions dealing with both air and space law, or at least some provisions of those conventions, may have achieved the status of customary international law. To the extent that customary international law has evolved from these conventions, it will be necessary to consider how such convention-based custom, or any other relevant sources of customary international law, will or should affect any future civil space safety standards.

[41] For details of ICAO's activities, visit: http://www.icao.int/ (last accessed: 03 January 2011).

[42] Convention on International Civil Aviation, 7 Dec. 1944, 15 U.N.T.S. 295 (hereinafter referred to as the Chicago Convention). For key provisions of the Convention, see Appendix B to this Study.

[43] For details, see Paul Stephen Dempsey and Michael Mineiro, "The ICAO's Legal Authority to Regulate Aerospace Vehicles," in Joseph N. Pelton and Ram S. Jakhu (eds.), Space Safety Regulations and Standards, 2010, Elsevier, 245, *et seq.*

[44] http://www.icao.int/icao/en/trivia/peltrgFAQ.htm (last accessed; 03 January 2011). For details see, Making An ICAO Standard, http://www.icao.int/icao/en/anb/mais/#1 (last accessed: 03 January 2011).

[45] http://www.icao.int/icao/en/trivia/peltrgFAQ.htm (last accessed; 03 January 2011).

[46] Two of the 18 Annexes to the Chicago Convention do not deal with technical issues. The SARPs contained in these Annexes, which address facilitation and security, are handled by the Air Transport Bureau.

[47] Continuation and Expansion of the ICAO Universal Safety Oversight Audit Programme, 2336th Report to Council by the President of the Air Navigation Commission, ICAO Doc. C-WP/11252 (of 29 November 1999); online: http://www.icao.int/Hyperdocs/display.cfm?V=2&name=C-WP%2F11252&Lang=E (last accessed: 03 January 2011).

[48] http://www2.icao.int/en/AVSEC/USAP/default.aspx (last accessed: 03 January 2011).

[49] Assad Kotaite. *Formal Regulatory Framework Needed to Govern Expanding Operations in Outer Space*, 55 ICAO J. 5 (2000); Assad Kotaite, *Space for new regulations*, FLIGHT SAFETY AUSTRALIA, March–April, 2001, p. 58.

[50] "Concept of Suborbital Flights: Information from the International Civil Aviation Organization (ICAO)", Committee on the Peaceful Uses of Outer Space Legal Sub-committee, 49th session, 2010, UN Doc. A/AC.105/C.2/2010/CRP.9 (19 March 2010). Also see: Peter van Fenema, *Suborbital Flights and ICAO*, Air & Space Law, Vol. XXX/1 (November 2005), pp. 396 *et seq.; Ruwantissa Abeyratne*, *ICAO's Involvement In Outer Space Affairs – A Need For Closer Scrutiny?*, Journal of Space Law, *Vol.* 30 (2004), pp. 185 *et seq.*

[51] ITU, online: http://www.itu.int/net/about/ (last accessed: 03 January 2011). For key provisions of the ITU Constitution, Convention and Radio Regulations, see Appendix A to this Study.

[52] ITU, online: http://www.itu.int/net/about/membership.aspx (last accessed: 03 January 2011).

[53] ITU, online: http://www.itu.int/osg/spu/ip/chapter_one.html (last accessed: 03 January 2011).

[54] ITU, online: http://www.itu.int/net/about/itu-t.aspx (last accessed: 03 January 2011).

[55] ITU, online: http://www.itu.int/dms_pub/itu-d/opb/gen/D-GEN-OVW-2007-E09-PDF-E.pdf (last accessed: 03 January 2011).

[56] IMO, online: http://www.imo.org/About/Pages/Default.aspx (last accessed: 03 January 2011).

[57] Convention on the International Maritime Organization, 6 Mar. 1948, 289 U.N.T.S. 48 (entered into force 17 Mar. 1958) [IMO Convention]. The IMO Convention was formerly known as the "Convention on the Inter-Governmental Maritime Consultative Organization" and the IMO was originally known as the International Maritime Consultative Organisation. Its name was officially changed in 1982.

[58] There is a Sub-committee dealing with each of the following subject areas: Bulk Liquids and Gases; Carriage of Dangerous Goods, Solid Cargoes and Containers; Fire Protection; Radio-communications and Search and Rescue; Safety of Navigation; Ship Design and Equipment; Stability and Load Lines and Fishing Vessels Safety; Standards of Training and Watch Keeping; Flag State Implementation.

[59] IMO, online: http://www.imo.org/About/Pages/Structure.aspx (last accessed: 03 January 2011).

[60] There are some instances in which customary international law would not be applicable to a particular State, such as in cases where that State was a persistent objector to the customary law at issue.

[61] Trail Smelter Arbitration (1949) 3 Review of International Arbitration Awards, 1965–1966.

[62] The International Court of Justice has discussed the emergence and requirements for customary international law in its opinions, including North Sea Continental Shelf (F.R.G. v. Den/F.R.G. v. Neth., 1969 I.C.J. 3, 41 at ¶ 88 (20 February 1969) and Legality of the Threat or Use of Nuclear Weapons (1996), available at http://www.icj-cij.org/docket/index.php?p1=3&p2=2&PHPSESSID =4ee804a21276b739255cdae42efca1fe (last accessed: 03 January 2011).

2.3 Transition from air law and space law to aerospace law

2.3.1 Introduction

Both the existing regimes of air law and of space law were developed at a time when the technology for Earth-to-Earth aerospace movements did not yet exist. Thus, there is not yet a unified or integrated regime of aerospace law, and there appears to be much overlap and inconsistency between the regimes of air law and space law. At the outset, one must determine which regime applies – air law, space law, or in some instances, both – and then identify the governing rules. The international legal regime governing air transport on issues such as liability, security, navigation, and air traffic management are well developed, and set forth in various conventions, treaties, and various "soft law" standards. Five multilateral space law instruments also define legal rights and duties. Yet at present, it is unclear whether aerospace vehicles fall under established principles of air law, and if they do, whether these laws follow them into space. Moreover, it is unclear where the legal limits of airspace expire, and the regime of outer space begins, and vice versa.

Outer space has no beginning and no end. While not knowing the end is clearly a matter of the limits of our current scientific knowledge, not having yet defined the beginning is just a sign of international neglect for the global commons of space.

The atmosphere is a relatively thin layer of gas surrounding the Earth, kept in place by the Earth's gravitational field, whose density decreases with height. Several "soft" boundaries between air and space have been defined: around 50 km is the upper limit of atmospheric buoyancy (weather balloons); 80 km is the threshold altitude that defines "astronauts" in the U.S., 100 km, also known as the "Karman Line", is where aircraft aerodynamic controls become ineffective; 108 km is the highest altitude achieved by an experimental aircraft; 120 km begins the re-entry threshold for space vehicles; and, 160 km is the lowest practical operating orbit for satellites and spacecraft. For the purpose of space exploration and use, 160 km is considered to be the most functional working definition.

Although, the 100 km separation between the field of aeronautics and that of astronautics has been recognized for the application of national space-related regulations by some countries such as Australia, currently there is no legally defined boundary mentioned in international aeronautical conventions and space treaties.

In fact, pursuant to Article 1 of the Chicago Convention, "the contracting States recognize that every State has complete and exclusive sovereignty over the airspace above its territory". Yet, there is no definition of the height of this territorial airspace. On the other hand, under Article I of the Outer Space Treaty, "Outer space . . . shall be free for exploration and use by all States without discrimination of any kind, on a basis of equality and in accordance with international law, and there shall be free access to all areas of celestial bodies". Moreover, according to Article II of the Outer Space Treaty, "Outer Space . . . is not subject to national appropriation by claim of sovereignty, by means of use or occupation, or by any other means". There is no indication in those instruments concerning any vertical limit of the airspace from where outer space would begin. COPUOS has discussed the issue of the definition and delimitation of outer space since 1962; however, no definite conclusion has been reached so far.[63]

One way to answer the question as to which regime of law applies is to ask what type of vehicle is being considered – is it an aircraft, or a spacecraft, or an aerospace vehicle? This is the *functionalist* approach to the problem. Another way is to ask where the object at issue is – is it in airspace, or in outer space, or does it traverse both? This is the *spatialist* approach to the question. These two approaches are discussed in detail under Sects. 2.3.3.1 and 2.3.3.2 *infra.*

The principal multilateral Conventions contain elements of both functionalism and spatialism. The Chicago Convention of 1944 applies principles of air law to "*airspace*" and to "*aircraft*". The Outer Space Treaty of 1967 applies principles of space law to "*outer space*" and "*space objects*". Unfortunately, in neither of these Conventions are those essential terms defined. These definitional failures create uncertainty and potential conflict between these two, quite different, legal regimes.[64]

In instances where both regimes apply, there will be a certain amount of inevitable inconsistency. As commercial aerospace flights become more common, their use of airspace also traversed by aircraft will proliferate, creating the need for defined rules of safety, security, and liability.

Before we discuss the legal aspects of the problem of boundary between airspace and outer space, it will be useful to understand the legal principles of the international regulatory regime established in several space law conventions.

2.3.2 Space law conventions

With the launch of Sputnik in 1957, space was initially the province of the Cold War superpowers. Concerned about the extension of "national rivalries" into space, seeking to promote the full "exploration and exploitation of outer space for the benefit of mankind", and attempting to reserve "the study and utilization of space for peaceful purposes", the United Nations created COPUOS the following year.[65] In a relatively brief period – from 1967 to 1979 – COPUOS drafted the following five international agreements and conventions:

- The Treaty on Principles Governing the Activities of States in the Exploration and Use of Outer Space, including the Moon and Other Celestial Bodies (the "Outer Space Treaty"),[66] which establishes the basic framework of international law applicable to space, declaring space to be the "province of all mankind". However, several important concepts are nowhere defined in the convention, such as "outer space" and "space object". The Outer Space Treaty provides that the exploration and use of outer space shall be the province of all mankind, free for exploration and use by all States, and not be subject to national appropriation by claim of sovereignty, by use or occupation, or by any other means. States shall be responsible for national space activities whether carried out by governmental or non-governmental entities, and shall be liable for damage caused by their space objects.
- The Agreement on the Rescue of Astronauts, the Return of Astronauts and the Return of Objects Launched into Outer Space (the "Rescue Agreement"),[67] which requires that States take all possible steps to rescue and assist astronauts in distress, and to return them to the launching State, and upon request, to provide assistance to launching States in recovering space objects and their component parts that fall back to Earth.
- The Convention on International Liability for Damage Caused by Space Objects (the "Liability Convention"),[68] which imposes absolute liability upon a launching State to pay compensation for personal injury and property damage caused by its space objects on the surface of the Earth, or to aircraft in flight.[69]
- The Convention on Registration of Objects Launched into Outer Space (the "Registration Convention"),[70] which requires each launching State to maintain a registry of objects they launch into space, and to furnish to the U.N., as soon as practicable, certain specified information concerning each space object.
- The Agreement Governing the Activities of States on the Moon and Other Celestial Bodies (the "Moon Agreement"),[71] which provides that the Moon and other celestial bodies should be used exclusively for peaceful purposes, its environment should not be disrupted, and that its natural resources must be shared with developing nations in some, as yet, undefined way.

As can be seen from the foregoing, these space treaties provide generic principles but very few specific and detailed implementing rules. They were produced at a time when the U.S. and the Soviet Union, then the two leading countries in space exploration, were locked in the Cold War. They have had the effect of entrenching the initial governmental monopoly on space activities which had resulted in a minimal presence of the private and commercial sector in space activities. The space treaties were, therefore, primarily and largely conceived for the purpose of defining the overall limits applicable to a nation's space activities. They were not designed to specifically and clearly facilitate and promote commercial and civil exploration and exploitation of outer space.

During the negotiation and drafting of the Outer Space Treaty in the late 1960s, there was some initial disagreement regarding the legal status of space activities carried out by the private sector. The United States wanted to leave the door open to private sector involvement in future space exploitation. On the contrary, the Soviet Union opposed this idea to the extent that the draft they proposed included the following statement: "All activities of any kind pertaining to the exploration of outer space shall be carried out solely and exclusively by States . . . ."

The United States then proposed a compromise solution that represents a fundamental difference in space law as compared to both maritime and air law. Each country would bear responsibility for the activities of its nationals in space. This compromise was accepted by the Russians and was subsequently incorporated into Article VI of the Outer Space Treaty in the following language: "States . . . shall bear international responsibility for national activities in outer space . . . whether such activities are carried on by government agencies or by non-governmental entities, and for assuring that national activities are carried out in conformity with . . . [this] Treaty. The activities of non-governmental entities in outer space . . . shall require authorization and continuing supervision by the appropriate State party to the Treaty".

Whereas under the space treaties, States are both "responsible" and "liable" for the space activities of their nationals (persons, companies, entities, etc.), they only exercise a supervisory role (responsibility) in connection with their commercial ships and aircraft and they do not bear financial risk (liability) unless it is determined that a genuine link exists between the concerned ship/aircraft and the State.

According to some interpretations, Article VI would prohibit strictly private, unregulated (i.e., unauthorized) activity in space. The terms "authorization" and "continuing supervision", therefore, imply licensing and enforced adherence to government imposed regulations by each State party involved. In stark contrast to maritime and air law, the UN space treaties make countries both "*responsible*" and "*liable*" for the space activities of their nationals (i.e., individuals, companies, entities, etc.). As such, under existing international space law, governments have both supervisory and financial responsibility for all space activities.

Furthermore, another UN space treaty, the 1972 Liability Convention affirms Article VII of the Outer Space Treaty by designating the launching State as the sole entity which is liable for damage. Moreover, Article II of the Convention stipulates that "a launching State shall be absolutely liable to pay compensation for damage caused by its space object on the surface of the Earth or to aircraft in flight". As a result, liability attaches primarily to the country that launches, procures a launch, or from whose territory or facilities a launch takes place.

Finally, in relation to activities that could "cause potentially harmful interference with activities of other States", a State is obliged under Article IX of the Outer Space Treaty to "undertake appropriate international consultations before proceeding with any such activity". Thus, Article IX does require safety cooperation in the form of undefined consultation in the event that the proposed space activities of one state pose a risk of interference with the space activities of other states. However, in practice, such consultations have basically consisted of participation in a few specialized committees such as COPUOS, which issues only generic guidelines, if any at all.

2.3.3 Boundary between airspace and outer space

Freedom of the use of space is a fundamental principle of the space law regime. In 1960, Julian Verplaetse observed that so long as the use of space is limited by technological and financial inhibition, freedom does not interfere with mutual activities: "But, once many States would join in that activity, some technical understanding and perhaps some legal agreement would be necessary. When the activity goes beyond its limited scope and becomes a usual practice, the need will be urgent and unabatable".[72] That need is rapidly approaching. As one source notes, "The number of commercial space vehicle developers and spaceport operators is sufficiently large – and the number continues to grow – that some form of structured control and regulation appears fully justified and necessary".[73] Patricia Smith, former FAA Associate Administrator for Commercial Space Transportation notes that since that office was created in 1984, it has licensed 181 launches and six commercial launch sites. She continued, "the time will come when we have many more vehicles flying in the national airspace with far greater regularity, launching from various launch sites, with passengers going point-to-point, that a regulatory approach that is a variant to the aviation certification process may be called for".[74]

In view of human space flight (including for space tourism), the definition and delimitation of outer space, which is the oldest item on the agenda of COPUOS, is a crucial legal problem. Yet, COPUOS has been unable to reach any consensus on the issue, and moreover, has been unable to promulgate any space law treaty or convention over the last three decades.

2.3.3.1 Spatialist approach: precisely, where is it?

One way of determining what law applies is to assess the location of the object. For example, an amphibious vehicle may be subject to the Law of the Sea when upon the high seas, and subject to the laws governing land transport when it comes ashore. In *Reinhardt v. Newport Flying Service Corp.*,[75] Judge Cardozo, speaking for a unanimous court, held that a hydroplane, moored and anchored in navigable waters, was a "vessel" within admiralty jurisdiction, rather than an aircraft. But Cardozo was careful to point out that even a hydroplane, while in the air, is not subject to the laws of admiralty. Under the same reasoning, an aerospace vehicle might be considered a spacecraft while in space, and an aircraft while in airspace.

Yet, the two regimes could not be more different. Airspace over national territory is, under the provisions of the Chicago Convention, subject to complete and exclusive State sovereignty, while under the *Outer Space Treaty*, States are explicitly prohibited from exercising any form of territorial sovereignty in space. Manfred Lachs, of the International Court of Justice, has observed: "*Cuius est solum eius est usque ad colum et ad sidera*; this principle, inherited from Roman Law, was accepted by international law and adapted to its needs. For centuries States faced no practical questions concerning the control over their airspace, nor the height to which their sovereignty extended .... Only the first journey of a man-made satellite reopened the issue: what was *coelum*? Can sovereignty extend *ad infinitum*?"[76]

The reason the air law and space law regimes approach spatial issues so differently is rooted in defence and military considerations. The Chicago Convention, and its predecessor, the Paris Convention of 1919, were both drafted after aircraft technology had revealed its destructive capacity in war. Thus, exclusive sovereign territoriality was deemed essential to protect the State from attack. The *Outer Space Treaty* was concluded during the Cold War – and in the midst of a very "hot war" in Vietnam – amid the anxiety that space might become a new arena for conflict between the superpowers. By denying sovereignty in space, the major powers sought to diffuse potential conflict. The commercial use of space, which would be realized decades hence, likely was not seriously considered during its negotiation in the mid-1960s, before man set foot on the Moon.

2.3.3.1.1 Territorial airspace

Article 1 of the Chicago Convention of 1944 reaffirms Article 1 of the Paris Convention of 1919, by recognizing the pre-existing rule of customary international law, that "every State has complete and exclusive sovereignty over the airspace above its territory". Territory is defined by Article 2 of the Chicago Convention as "the land areas and territorial waters adjacent thereto under the sovereignty, suzerainty, protection or mandate of each State". Thus, an object flying through territorial airspace would fall under the domestic aviation laws of the underlying State. Under Chicago's Article 6, scheduled international air services may not enter the airspace of a State without its permission, and subject to any conditions the State may impose.[77] Multilaterally, many States have exchanged reciprocal transit (First and Second Freedom) rights through the Transit Agreement.[78] Commercial rights are usually exchanged through bilateral air transport agreements.

Thus, there is no corresponding right in aviation law to the maritime law concept of "freedom of the seas" (as was originally proposed by Hugo Grotius as early as 1609) or the right of "innocent passage" through airspace over land and territorial waters, though there is the right to freely fly over the high seas.[79] Though a maritime vessel flying the flag of a non-belligerent State could freely participate in international trade and commerce at any seaport, an aircraft could not land at a foreign airport without that State's permission, nor could it take on or discharge passengers or cargo without its permission. Article 3 of the Law of the Sea Convention extends the jurisdiction of coastal States to 12 miles, while Article 38 establishes a right of transit in the straits for military and commercial aircraft.[80]

Article 96 of the Chicago Convention defines an *international air service* as "an air service which passes through the airspace over the territory of more than one State". An *air service* is defined as a scheduled service performed by aircraft for the movement of passengers or property.[81] However, the term *aircraft* is undefined in the Convention. "The concept of atmosphere may help to define aircraft", observes Professor Bin Cheng. "But airspace is not limited to the use of aircraft since it includes all space where any wisp of air is found, even if it is insufficient to give support to aircraft."[82]

To ensure uniformity of rules of the air, under the Chicago Convention, States are obliged to adopt domestic aviation laws that conform to the Standards and Recommended Practices [SARPs] promulgated by ICAO, and included as Annexes to the Convention.[83] States are also obliged to ensure that aircraft flying over their territory or carrying their nationality mark observe the rules and regulations governing flight and navigation there in force.[84] For example, the

United States has promulgated laws governing commercial space launches, vehicles, crew, and navigation, vesting jurisdiction in its Federal Aviation Administration [FAA], which has comprehensive jurisdiction over aircraft and aviation safety and navigation.[85] The German Federal Aviation Code also specifies that "spacecraft, rockets and similar flying objects" are considered to be aircraft while in airspace, and thus subject to the prevailing rules and regulations governing aircraft.[86]

The right of innocent passage through territorial airspace for ascending or descending space objects has not been established under either conventional or customary international law. The U.S. Space Shuttle usually ascends and descends over U.S. airspace or over the oceans. On relatively few occasions has a space object flown over the territorial airspace of a State other than the launching State; when territorial airspace has been entered, the reason for the absence of objection usually is because the underlying State was unaware of the territorial intrusion, not because it acquiesced in the breach of its territorial sovereignty over its airspace. On other occasions, entry was requested and granted, as in 1990 when the Soviet Union granted the United States permission to overfly its territory on the final flight stage of the Atlantis shuttle.[87]

2.3.3.1.2 Airspace above the high seas

Airspace beyond the territorial seas, and above the high seas, is open for use by all. However, under the Chicago Convention, the rules governing such airspace are those promulgated by ICAO in SARPs (i.e. Annexes to the Chicago Convention).[88] It is interesting to note that Appendix B provides that no State can file difference with ICAO rules over the high seas adopted in accordance with Article 12 of the Chicago Convention. Hence, there is strong precedent for ICAO to exert jurisdiction over aviation in non-sovereign space. As noted above, ICAO has jurisdiction to promulgate rules of safety and navigation which are binding to flights covering 72% of the Earth's surface.

2.3.3.1.3 Outer space

The *Outer Space Treaty* provides that the "exploration and use of outer space . . . shall be the province of all mankind".[89] It declares outer space to be the common property of mankind, to be used freely "for exploration and use by all States",[90] and not to be subjected to national appropriation or otherwise subjected to the sovereignty of any State.[91] Thus, outer space is free for use by all.

Certain equatorial States have attempted to assert sovereignty over the geostationary orbit above their territories.[92] But these declarations have not been recognized by other States on grounds that such territorial claims are inconsistent with Articles I and II of the *Outer Space Treaty of 1967*.[93]

No rule of conventional or customary international law defines where airspace ends and outer space begins. Some have suggested the Karman Line be adopted as the dividing line between airspace and outer space, approximately 100 km above the Earth's surface.[94] Certain national laws, such as the legislation of Australia, define outer space as that above 100 km. Although the issue has been debated since 1962, COPUOS has reached no consensus.[95] Arguably, both orbital and suborbital flights in space, though enjoying the right of free transit, would fall under space law.[96]

2.3.3.1.4 Problems with the spatialist approach

The spatialist approach poses many problems. One is that there is no consensus as to where to draw the line of demarcation between airspace and outer space. Professor Bin Cheng has observed, "there are probably as many criteria as there are speakers and writers on the subject: gravitational effect, effective control, actual lowest perigee of orbiting satellites, theoretical lowest perigee of orbiting satellites, the von Karman line, limit of air drag, limit of air flight, the atmosphere and its various levels, an absolutely arbitrary height ... (100 km) or one-hundredth of the Earth's radius (64 km...)".[97] In the early, 20th century, some scholars proposed borrowing from maritime law and limiting territorial airspace according to the cannon-shot rule.[98] Yet, without a demarcation of the boundary between airspace and outer space, the legal regime remains murky, with different States claiming sovereignty in areas that other States consider outer space, thereby creating potential conflict.[99]

Though both the Chicago Convention of 1944 and its predecessor, the Paris Convention of 1919, recognized that States enjoy complete and exclusive sovereignty over the airspace above their territories, neither instrument prescribed how high such a territorial claim could be made. Some also argue that establishing a boundary too high might hamper certain space activities; once established, it would be difficult to amend it, particularly if it should be lowered.[100] Moreover, if a legal question arose during a flight near the point of demarcation between airspace and outer space, it might be difficult to determine on which side of the line the event occurred.[101]

Another problem is that an aerospace vehicle may enter suborbital space for only a short time, while its primary activity and mission occurs in the airspace. Thus, it may be more appropriate to apply air law to the entire movement. A spatialist approach might require that vehicles be certified under, and regulated by, two separate legal regimes – one, an air law regime created by ICAO, and another, a space law regime created by some future space navigation organization.

2.3.3.2 Functionalist approach: precisely, what is it?

The other approach to the question of which legal rules govern the movement is to examine what kind of object is in question. Is the vehicle in question an aircraft, a space object, or an aerospace object?

There are several ways to answer this question. One is to identify the vehicle's purpose, its activity, or its destination. Is the vehicle's primary purpose to go into outer space for purposes of orbit, to conduct outer space activities (an Earth-Space mission), or to provide transportation from one point on Earth to another (an Earth-to-Earth mission)? In the former case, arguably it is a space object, while in the latter case, it may not be.

Another approach to answering this question is to consider the technological properties, functional characteristics, design, and aerodynamics of the vehicle. Is it capable of acquiring lift through the air, for example? Once it is categorized as a space object, arguably space law would apply to it throughout its journey; once categorized as an aircraft, then perhaps air law would apply, irrespective of its location.

2.3.3.2.1 Aircraft

If it is an *aircraft*, air law applies to it. The Chicago Convention of 1944,[102] – the *Magna Carta* of public international air law – applies to "civil aircraft", but does not define what is contemplated by the term "aircraft". Several years after the Paris Convention of 1919 entered into force, the International Commission for Air Navigation [ICAN] (the international organization created by that Convention), adopted a "Glossary of terms used in Aeronautical Technology". In that Glossary, the term "aircraft" was defined as "a machine which can derive support in the atmosphere from reactions of the air". In 1930, ICAN adopted a formal resolution providing for new definitions to be applicable to all the Annexes.[103] The types of aircraft described therein required full atmospheric support in order to navigate successfully. No means of propulsion then existed, or was then contemplated (beyond perhaps Jules Verne and other science fiction writers), by which any vehicle could have been navigated beyond the height where reactions of the gaseous air could give it full support and enable flight. During the several decades that the Paris Convention governed international aviation, the final definition of "aircraft" as a machine "which can derive support in the atmosphere from reactions of the air" necessarily described flight instrumentalities that required full aerodynamic support.[104] The U.S. definition under its domestic law – the Air Commerce Act of 1926 – was far more expansive in scope, defining an aircraft as "any contrivance now known or hereafter invented, used or designed for navigation or flight in the air."[105]

In 1967, well before commercial space transportation was feasible, the International Civil Aviation Organization amended the language it had borrowed from ICAN and incorporated within its Annex 7 to define an *aircraft* as "Any machine that can derive support in the atmosphere from the reactions of the air other than the reactions of the air against the Earth's surface."[106] This revised definition was aimed at making it clear that all air-cushion-type vehicles, such as hovercraft and other ground-effect machines, should not be classified as aircraft.[107] At the time, no thought apparently was given to the issue of whether an aerospace craft should be placed under ICAO's jurisdiction. All space launches in that era were State vehicles, and under the Chicago Convention State aircraft were exempt from ICAO regulation; hence ICAO did not have the capability then to exert jurisdiction over the Soviet Sputnik and its NASA progeny.

Under the current definition, an aerospace vehicle launched by rocket would not be considered an aircraft on the ascent phase of its flight, but might well on the descent phase, as it will be using its wings to glide to its destination. ICAO has the authority to amend Annex 7 to include aerospace vehicles within its definition, but as yet, has failed to do so. ICAO has recognized that, "Should suborbital vehicles be considered (primarily) as aircraft, when engaged in international air navigation, consequences would follow under the Chicago Convention, mainly in terms of registration, airworthiness certification, pilot licensing and operational requirements (unless they are otherwise classified as State aircraft under Article 3 of the Convention)".[108]

NASA's Space Shuttle was the first reusable orbital launch vehicle. The Chicago Convention exempts State aircraft from its scope.[109] Hence, the Shuttle would fall outside its scope. However, the Convention provides that when issuing regulations for State aircraft, due regard must be given to the navigational safety of civil aircraft.[110]

Given that aerospace vehicles use the same airspace as aircraft, at least for a period of time, it would seem desirable to apply a single unified regime of air navigation to both. Moreover, ICAO's 18 Annexes to the Chicago Convention governing issues such as safety, airworthiness, navigation, licensing, and communications would seem appropriate to govern both aircraft and aerospace vehicles occupying common airspace, rather than reinventing the wheel and crafting wholly new rules to govern only aerospace vehicles.[111] At this writing, the ICAO Council was evaluating the question of whether it should exert jurisdiction over suborbital flight.[112]

2.3.3.2.2 Space objects

If the space transportation vehicle is a "*space object*", space law presumably applies to it. However, none of the five space law conventions[113] define precisely what is contemplated by the term "space object",[114] and none were drafted with any thought being given to commercial space transportation.[115] Presumably, a spacecraft should be capable of moving in outer space (either orbital or suborbital) without any support from the air, and would have a power source not dependent upon external air or oxygen.

2.3.3.2.3 Aerospace vehicles

What if the space transportation vehicle is a hybrid *aerospace object*, one capable of achieving lift and thereby flying in airspace (on ascent, descent or both), as well as travelling in outer space? Thus, a vehicle like the NASA Space Shuttle might be considered a *space object* during its launch and ascent supported by rockets, and during the weightless portion of its flight through space, then an *aircraft* during its descent and landing phases.

It is likely that parts of air law and space law both apply to such an *aerospace object*. Some rules of space law would apply from launch to landing, while some rules of the air law would apply during the time the object is in airspace.

2.3.3.2.4 Problems with the functionalist approach

Many scholars opt for a functional approach that places the dividing line between airspace and outer space at approximately 100 km above the Earth, as the air is not adequate to support flight above 80 km, and satellites in orbit do not begin to fall back into the Earth's atmosphere below 120 km. The government of Australia has promulgated a domestic law demarcating space as beginning from 100 km above the Earth. But as yet, there is no conventional or customary international law defining the boundary between airspace and outer space. Professor Bin Cheng observes:

> The territorial scope of a State's jurisdiction extends upward into space and downwards to the centre [sic] of the Earth, the wheel in the shape of an inverted cone. The sides of each cone are formed by straight lines projected downwards to the centre of the Earth and upwards into space. As the Earth is very nearly a true sphere, there will be neither gaps nor overlapping in between these inverted cones.[116]

Under the functionalist approach, a suborbital vehicle that merely passes through outer space in the course of Earth-to-Earth transportation would presumably remain subject to air law during the entire journey; but a vehicle that passes through airspace in the course of Earth-to-Space transportation would be considered to be governed by space law during its entire journey. As ICAO has noted, "it might be argued from a functionalist viewpoint that air law would prevail since airspace would be the main centre of activities for suborbital vehicles in the course of an Earth-to-Earth transportation, any crossing of outer space being brief and only incidental to the flight".[117]

Because aerospace vehicles share airspace with commercial aircraft, the rules of air safety and navigation must be harmonious. If they operate under two separate legal regimes, the danger of aircraft and aerospace vehicle collisions increases. In one sense, the spatialist approach offers greater certainty as to the applicable legal regime, particularly on the rules of navigation governing commonly used airspace. Yet in another, demarcating where airspace ends and outer space begins has confounded both the scientific and legal experts for decades.

2.3.4 Need for a unified legal regime

As yet, it is unclear whether the Chicago Convention applies to suborbital and orbital commercial launch vehicles that transport passengers or freight internationally. The Convention explicitly excludes application to "State aircraft", so that NASA or ESA launches, for example, do not fall under its provisions.[118] The Chicago Convention does, however, apply to "civil aircraft".[119] But it is unclear whether a commercial aerospace vehicle constitutes a civil aircraft.

During the ballistic portions of its flight, while not supported by the reactions of the air, a spacecraft would not fall under this peculiar definition. However, upon descent, after it has re-entered the Earth's atmosphere and as it is gliding on its return to Earth, it could be considered an aircraft in flight, and therefore subject to the Chicago Convention.[120] Article 96(b) defines "international air service" as that which passes through the airspace over the territory of more than a single State.

A spacecraft launch or descent may constitute such an international air service as it traverses the airspace of different nations.

As noted above, Article 1 of the Chicago Convention provides, "The Contracting States recognize that every State has complete and exclusive sovereignty over the airspace above its territory".[121] This is a three-dimensional concept, recognizing that a State has sovereignty over its surface territory and territorial waters, as well as the airspace above it. But the Chicago Convention does not define the upper limit of airspace. Professor John Cobb Cooper wrote of "a universally accepted rule of international law that airspace above national lands, waters, and territorial waters is part of the territory of the subjacent State, and that each sovereign State has the same right to control all movements in its national airspace as it had on national lands and waters . . . ".[122]

However, under the Outer Space Treaty,[123] no nation may expropriate space, and hence, no national jurisdiction can be exerted in it. Outer space is not subject to national appropriation by claim of sovereignty, by means of use or occupation, or by any other means. Yet, as discussed *supra*, the line of demarcation between airspace and outer space has not been legally drawn. The Earth's atmosphere does not end abruptly, but becomes progressively thinner with elevation. As the planet's gravity diminishes with elevation, objects gradually disappear into space.

Future transportation systems will be highly influenced by the legal regime in which they are developed. Commercial development of space would be greatly enhanced by clarity, stability, and predictability of law.[124] Lack of uniformity of law, and conflicting and overlapping laws, will impair the market's interest in investment in space transportation, and the insurance industry's ability to assess and price risk.[125]

Commercial investment in space transportation systems is expensive, depends on as yet unproved technology, and is fraught with risk. Clear legal rules can help define the degree, or consequences, of risk, and reduce uncertainty, providing the predictability necessary to support commercial investment. Conversely, legal uncertainty can increase risk and dampen enthusiasm for investment.

Many commentators have urged that legal rules be refined to take account of commercial needs in space.[126] Some have suggested that the emerging legal regime should be one of air law.[127] Others prefer the regime of space law.[128] Still others have urged immunity from liability for commercial activities in space for a developmental period.[129]

Probably the simplest, and most sensible initial effort would be for ICAO to amend its annexes to redefine aircraft to include aerospace vehicles, so that when they fly in the airspace used by civil aircraft, the rules of safety and navigation are the same.[130] It could do so by amending the definition of "aircraft" to include aerospace vehicles. It created the definition of aircraft, and amended it to clarify that air cushion vehicles were not within the Chicago Convention; ICAO could amend its Annexes again to clarify that suborbital vehicles fall within the definition of "aircraft". One potential definition that might be used as a model was that promulgated by the U.S. Congress in the Air Commerce Act of 1926: "any contrivance now known or hereafter invented, used or designed for navigation or flight in the air".[131] The Canadian Parliament has defined an aircraft as "any machine capable of deriving support in the atmosphere from reactions of the air, and includes a rocket".[132] Another source recommends that suborbital vehicles be included in the air law regime, and orbital vehicles be placed within the space law regime.[133] ICAO could amend particular Annexes to define the rules of safety and navigation for "aircraft" so redefined. As the organization that drafted the Montreal Convention of 1999 addressing air carrier liability, and the several aviation security conventions, ICAO could also clarify whether aerospace vehicles fall under their provisions as well.

Alternatively, ICAO could promulgate a new Annex on "Space Standards". There is precedent for this as well. Article 37 of the Chicago Convention vests in ICAO the authority to promulgate Standards and Recommended Practices as Annexes to the Convention. Therein, it lists eleven specific areas to which ICAO is instructed to devote itself, mostly focusing on safety and navigation. Yet, since its creation, as air transport has grown and evolved, ICAO has focused on other areas not explicitly listed in Article 37, including, for example, the promulgation of wholly new annexes addressing environmental and security issues. Article 37 is sufficiently broad to permit such jurisdictional assertions, as it provides that ICAO may promulgate SARPs addressing "such other matters concerned with the safety, regularity, and efficiency of air navigation as may from time to time appear appropriate".[134] A certain international regulatory body is needed to provide uniform standards for national certification of space launch systems and vehicles, and their navigation through airspace.

ICAO might also define the limits of airspace by amending an Annex, though some may argue that such a change would require a new protocol amending the Chicago Convention itself, or perhaps an entirely new multilateral convention. This is by no means a new proposal. As early as 1956, Professor John Cobb Cooper urged that the definition of airspace should be determined by the United Nations and that pertinent regulations should be promulgated by ICAO.[135]

Others may argue that a separate space traffic management system, under a new international space management organization, should be established. As early as 1960, Julian Verplaetse insisted, "It has been questioned whether ICAO should amend its Annexes and widen the scope of its definition of aircraft so as to include rockets and missiles and even satellites. In view of the specific character of outer space law and inasmuch as those contrivances are mostly used in Outer space, it is suggested that spacecraft are different from the contraptions regulated by the air law conventions and should be dealt with in separate international instruments".[136] Yet, these observations were written at a time when the only space activities were those launching satellites into orbit. Today, we confront the issue of suborbital vehicles, which are very similar to "contraptions regulated by the air law conventions". Moreover, it would be difficult to justify replication of the able and detailed work already done by ICAO on issues such as safety, navigation, security, and liability, at least with respect to flights in the Earth's atmosphere.

Dr. Nandasiri Jasentuliyana has called for COPUOS to promulgate "Space Standards" similar to ICAO's SARPs, and to draft a convention creating an international framework for space vehicles.[137] Yet, for three decades, COPUOS has been unable to promulgate any multilateral legal instrument for ratification by States.[138] If COPUOS is able to break its deadlock, so much the better. If not, as the United Nation's arm for air transportation, ICAO should provide clarification on the issues of what is contemplated by aircraft, and what is contemplated by airspace, and then proffer standards of harmonization as SARPs, which member States would be obliged to follow. Under the Chicago Convention, member States are obliged "to collaborate in securing the highest practicable degree of uniformity" on such issues,[139] and to "keep [their] own regulations . . . uniform, to the greatest possible extent", with SARPs.[140]

Formal clarification of what law applies would be highly desirable. The fundamental legal principles embedded in the respective legal regimes of air law and space law are quite different – one recognizing territorial sovereignty, and the other denying it; one imposing limited liability upon the carrier, and the other imposing unlimited liability on the State. These conflicts and inconsistencies may unravel the uniformity of law that the conventions seek to attain, and inhibit investment in commercial space transportation systems.

The time has come for the international community to promulgate conventional international law with an eye to facilitating – and indeed, promoting – commercial aerospace activities. Space transportation would also be facilitated by harmonizing space laws with the prevailing rules of safety, navigation, security, and liability applicable under air law; thus initiating a logical transition from air law and space law to aerospace law. The public's safety demands no less.

[63] For a detailed analysis of the international debate on the legal aspects of boundary between airspace and outer space, see Sects. 2.3.3 and 2.3.4 infra.

[64] This problem has been debated and discussed by legal scholars since the 1950s and 1960s, with no resolution.

[65] U.S. General Ass. Res. 1348, reproduced in Paul Stephen Dempsey, Space Law III.B1-1 (2008).
[66] Treaty on Principles Governing the Activities of States in the Exploration and Use of Outer Space, including the Moon and other Celestial Bodies (hereinafter referred to as the Outer Space Treaty); opened for signature on 27 January 1967; ratified by 100 States and signed by 26 additional States (as of 1 January 2010); 610 U.N.T.S. 205.
[67] The Agreement on the Rescue of Astronauts, the Return of Astronauts and the Return of Objects Launched into Outer Space, opened for signature on 22 April 1968, (hereinafter referred to as Rescue and Return Agreement); ratified by 91 States and signed by 24 additional States (as of 1 January 2010); 672 U.N.T.S. 119.
[68] The Convention on International Liability for Damage Caused by Space Objects (hereinafter referred to as the "Liability Convention"), opened for signature on 29 March 1972; ratified by 88 States and signed by 23 additional States (as of 1 January 2010); 961 U.N.T.S. 187.
[69] See generally, Marc S. Firestone, Problems in the Resolution of Disputes Concerning Damage Caused in Outer Space, 59 Tulane Law Review 747 (1985).
[70] The Convention on Registration of Objects Launched into Outer Space (hereinafter referred to as the "Registration Convention"), opened for signature on 14 January 1975, ratified by 53 States and signed by 4 additional States (as of 1 January 2010); 1023 U.N.T.S. 15.
[71] The Agreement Governing the Activities of States on the Moon and Other Celestial Bodies, (hereinafter referred to as the "Moon Agreement"); opened for signature on 18 December 1979; ratified by 13 States and signed by 4 additional States (as of 1 January 2010); UN General Assembly Resolution 34/68.
[72] Julian Verplaetse, International Law in Vertical Space, 153 (Rothman, 1960).
[73] George Washington University, SACRI Research Study 46 (2008).
[74] How Safe Is Space Tourism?, Wall Street Journal (19 April 2007).
[75] 133 N.E. 371 (N.Y. 1921).
[76] Manfred Lachs, The Law of Outer Space: An Experience in Contemporary Law-Making, 42 (1972).
[77] Chicago Convention, Article 6. Pilotless aircraft also may not be operated over a State's territory without its permission. Chicago Convention, Article 8. Non-scheduled flights, however, enjoy the right of transit for non-traffic purposes. Chicago Convention, Article 5. However, the State may establish prohibited areas for reasons of military necessity or public safety. Chicago Convention, Article 9.
[78] International Air Services Transit Agreement (IASTA), ICAO Doc. 7500 (7 December 1944) also known as the Two Freedoms Agreement. However, many geographically important States – including the Russian Federation, Canada, China, Brazil and Indonesia – have not ratified the Transit Agreement.
[79] Article 12 of the Chicago Convention imposes the standards created by ICAO on flights over the high seas. I. H. Ph. Diederiks-Verschoor, An Introduction to Air Law, 32 (6th ed. 1997). Moreover, Article 5 of the Chicago Convention conferred the right of non-scheduled flights to fly over the territory of another State for non-traffic purposes, though the underlying State could impose conditions on the overflight. An ICAO delegate to the Legal Sub-committee of COPUOS observed, "the right of innocent passage of spacecraft through sovereign airspace . . . does not exist under present international law of the air; an unconditional right of passage through the sovereign airspace does not exist even with respect to the civil aircraft and is specifically subject to a special authorization with respect to State aircraft and pilotless aircraft". ICAO Doc. C-WP/8158 of 15/1/86.
[80] I. H. Ph. Diederiks-Verschoor, An Introduction to Air Law, 33 (6th ed. 1997).
[81] Chicago Convention, Article 96.
[82] Quoted in Julian Verplaetse, International Law in Vertical Space, 146 (Rothman, 1960).
[83] Chicago Convention, Article 37.
[84] Chicago Convention, Article 12.
[85] See http://www.faa.gov/about/office_org/headquarters_offices/ast/ (last accessed: 03 January 2011).
[86] Comments of Germany in A/AC.105/635/Add. 11 (Jan. 26, 2005) in COPUOS, Compilation of Replies Received from Member States to the Questionnaire on Possible Legal Issues with Regard to Aerospace Objects, http://www.unoosa.org/oosa/en/SpaceLaw/aero/index.html (last accessed: 03 January 2011).

[87] Comments of the Russian Federation A/AC.105/635/Add. 1 (15 March 1996) in Compilation of Replies Received from Member States to the Questionnaire on Possible Legal Issues with Regard to Aerospace Objects in http://www.unoosa.org/oosa/en/SpaceLaw/aero/index.html (last accessed: 03 January 2011). Article 19 of the Russian Federation Act on Space Activity of 1993 authorizes a single innocent flight through its airspace provided sufficient notice of the time, location and flight path is conferred in advance.
[88] Chicago Convention, Article 12.
[89] Outer Space Treaty, Article 1.
[90] Outer Space Treaty, Article 1.
[91] Outer Space Treaty, Article 2.
[92] The 1967 Bogota Declaration sought to achieve sovereignty by equatorial States over the geostationary orbit above them. See Space Law I.B14.5 (P. Dempsey, ed. Oceana 2007).
[93] Ram Jakhu, "The Legal Status of the Geostationary Orbit", VII Annals of Air and Space Law, 333 (1982).
[94] Physicist Theodore von Karman calculated that this was the height at which a vehicle would have to travel at orbital velocity to have sufficient lift to derive support from the atmosphere. The air is not adequate to sustain flight at above approximately 80 km, and satellites begin to fall back into the Earth's atmospheres at about 120 km. See Dean N. Reinhardt, "The Vertical Limit of State Sovereignty", 72 Journal of Air Law and Commerce, 65 (2007).
[95] In fact, COPUOS has been unable to produce a treaty of any kind since the Moon Agreement in 1979.
[96] U.S. law defines a suborbital flight as: "The intentional flight path of a launch vehicle, re-entry vehicle, or any portion thereof, whose vacuum instantaneous impact point does not leave the surface of the Earth." 49 U.S.C. § 70102 (20).
[97] Bin Cheng, "The Legal Regime of Airspace and Outer Space: the Boundary Problem", V Annals of Air and Space Law, 323 (1980).
[98] Wybo P. Heere, "Problems of Jurisdiction in Air and Outer Space," XXIV Air and Space Law (1999).
[99] Varlin Visssepo, "Legal Aspects of Reusable Launch Vehicles," 31 Journal of Space Law 165, 175 (2005).
[100] Robert Goedhart, The Never Ending Dispute: Delimitation of Air Space and Outer Space (1996).
[101] Varlin Visssepo, "Legal Aspects of Reusable Launch Vehicles", 31 Journal of Space Law, 165, 172 (2005).
[102] Chicago Convention, Article 3(a).
[103] The old definitions quoted above disappeared from Annex D and the new definitions appeared at the head of Annex A, reading as follows:
"The terms used in Annexes A to G have the following meanings:
- The word 'aircraft' shall comprise all machines which can derive support in the atmosphere from reactions of the air;
- The word 'aerostat' shall mean an aircraft supported in the air statically;
- The word 'balloon' shall mean an aerostat (free or captive) non-mechanically-driven;
- The word 'airship' shall mean a mechanically-driven aerostat with means of directional control;
- The word 'aerodyne' shall mean an aircraft whose support in flight is derived dynamically from the reaction on surfaces in motion relative to the air
- The word 'aeroplane' shall mean a mechanically-driven aerodyne supported in flight by aerodynamic reactions on surfaces remaining fixed under the same conditions of flight;
- The term 'glider' means a non-mechanically-driven aerodyne supported in flight by aerodynamic reactions on surfaces remaining fixed under the same conditions of flight."

The formal adoption of these definitions constituted a formal amendment to the Convention.
[104] John Cobb Cooper, State Sovereignty and Flight: An Historical Analysis (unpublished manuscript 1967).
[105] John Cobb Cooper, The Chicago Convention and Outer Space (address before the American Rocket Society Conference on Space Flight (New York, 24 April 1962).

[106] Chicago Convention, Annex 7. Moreover, the word aeroplane was defined as, "A power-driven heavier-than-air aircraft, deriving its lift in flight chiefly from aerodynamic reactions on surfaces which remain fixed under given conditions of flight." See also, South Africa Aviation Act No. 74 of 1962 § 1.
[107] ICAO, "The Concept of Suborbital Flights", Working Paper C-WP/12436 (30/05/05). Available online at: http://www.icao.int/ICDB/HTML/English/Representative%20Bodies/Council/Working%20Papers%20by%20Session/175/C.175.WP.12436.EN/C.175.WP.12436.EN.HTM (last accessed: 03 January 2011), reprinted in Paul Stephen Dempsey, Space Law § III.B3-1 (2006). ICAO also defined an "aeroplane" as, "A power-driven heavier-than-air aircraft, deriving its lift in flight chiefly from aerodynamic reactions on surfaces which remain fixed under given conditions of flight." *Ibid.*
[108] Paul Stephen Dempsey, Space Law § III.B3-1 (2006).
[109] Chicago Convention, Article 3(a).
[110] Chicago Convention, Article 3(d).
[111] Ram Jakhu and Yaw Nyampong, Are the Current International Space Treaties Sufficient to Regulate Space Safety, and Establish Responsibility and Liability? (unpublished paper 2007).
[112] ICAO Assembly Resolution A29-11 provides that ICAO shall continue to be responsible for stating the position of civil aviation on all related outer space matters.
[113] These treaties are: (1) The Treaty on Principles Governing the Exploration and Use of Outer Space including the Moon and Other Celestial Bodies; (2) The Agreement on the Rescue of Astronauts, Return of Astronauts, and the Return of Objects Launched into Outer Space; (3) The Convention on International Liability for Damage Caused by Space Objects; (4) The Convention on the Registration of Objects Launched into Outer Space and (5) The Agreement Governing the Activities of States on the Moon and other Celestial Bodies.
[114] The Liability and Registration Conventions indicate that a "space object" includes its component parts and its launch vehicle and parts thereof.
[115] ICAO, "The Concept of Suborbital Flights", Working Paper C-WP/12436 (30/05/05). Available online at: http://www.icao.int/ICDB/HTML/English/Representative%20Bodies/Council/Working%20Papers%20by%20Session/175/C.175.WP.12436.EN/C.175.WP.12436.EN.HTM (last accessed: 03 January 2011).
[116] Bin Cheng, "The Legal Regime of Airspace and Outer Space: the Boundary Problem", V Annals of Air and Space Law, 323 (1980).
[117] Paul Stephen Dempsey, Space Law § III.B3-1 (2006).
[118] Chicago Convention, Article 3.
[119] *Ibid.*
[120] ICAO, Council Working Paper C-WP/12436 30/05/05 (2005).
[121] Chicago Convention, Article 1.
[122] John Cobb Cooper, "Backgrounds of International Public Air Law", 1 Yearbook of Air and Space Law, 3, 23 (1967).
[123] Outer Space Treaty, Article 2.
[124] See generally, Stephan Hobe, "Aerospace Vehicles: Questions of Registration, Liability and Institutions", XXIX Annals of Air and Space Law, 377 (2004).
[125] What is needed is a "secure framework of regulations and legal responsibility ... [to] encourage increased activities in the future." Peter Nesgos, "Commercial Space Transportation: A New Industry Emerges", XVI Annals of Air and Space Law, 393, 412 (1991).
[126] Henri Wassenbergh, "Access of Private Entities to Airspace and Outer Space", XXIV Annals of Air and Space Law, 311, 325 (1999); Henri Wassenbergh, "The Art of Regulating International Air and Space Transportation", XXIII Annals of Air and Space Law, 201 (1998); Bruce Stockfish, "Space Transportation and the Need for New International Legal and Institutional regime", XVII-II Annals of Air and Space Law, 323 (1992).
[127] Charity Ryabinkin, "Let There Be Flight: It's Time to Reform the Regulation of Commercial Space Travel", 69 Journal of Air Law and Commerce, 101 (2004).
[128] Steven Freeland, "Up, Up and... Back: The Emergence of Space Tourism and Its Impact on the International Law of Outer Space", 6 Chicago Journal of International Law, 1 (2005). Blending

functionalist and spatialist principles, Prof. Freeland argues, "the most appropriate approach seems to be the application of space law... to the entire journey on the basis of the proposed function of the spacecraft carrying tourists – that is, the intention that it involves flight in outer space. The alternate 'exclusive' approach – to apply air law to the entire space tourism activity – appears unworkable given the lack of sovereignty that exists in outer space." *Ibid* at 9. Prof. Hobe makes a similar argument: "the provisions of the Chicago Convention are based on the principle of sovereignty in national airspace and are therefore generally not applicable to activities which take place in outer space". Stephan Hobe, "Aerospace Vehicles: Questions of Registration, Liability and Institutions", XXIX Annals of Air and Space Law, 377 (2004). Similarly, Prof. Zhao argues, "The air transportation regime, characterized by state sovereignty over airspace, substantially differs from the space travel regime.... This fundamental difference justifies the necessity of developing a distinct legal regime for space travel". Yun Zhao, "Developing a Legal Regime for Space Tourism: Pioneering a Legal Framework for Space Commercialization" (American Institute of Aeronautics and Astronautics 2005).

It is unclear why it is unworkable to have an Air Law regime apply to non-territorial outer space, inasmuch as a sophisticated body of both Public and Private International Air Law has developed involving intercontinental flights over the high seas, where no state has sovereignty. Over the high seas, which comprise more than 70% of the planet, the rules of the air are those established by ICAO. See Chicago Convention, Article 12.

[129] Susan Trepczynski, "The Benefits of Granting Immunity to Private Companies Involved in Commercial Space Ventures", XXXI Annals of Air and Space Law, 381, 403 (2006).

[130] Ram Jakhu and Yaw Nyampong, "Are the Current International Space Treaties Sufficient to Regulate Space Safety, and Establish Responsibility and Liability?" (unpublished paper 2007).

[131] John Cobb Cooper, The Chicago Convention and Outer Space (address before the American Rocket Society Conference on Space Flight (New York, 24 April 1962).

[132] Aeronautics Act, R.S.C. § 3(1) (1985).

[133] Varlin Vissepo, "Legal Aspects of Reusable Launch Vehicles", 31 Journal of Air Law and Commerce, 165, 214 (2005).

[134] Chicago Convention, Article 37.

[135] John Cobb Cooper, Legal Problems of Upper Space (address before the American Society of International Law, Washington, D.C. (26 April 1956), quoted in Andrew Haley, "The Law of Space – Scientific and Technical Considerations", 4 N.Y. L. Forum, 266 (1958). In 1962, Professor Cooper wrote that ICAO should interpret what is contemplated by "airspace" under Article I of the Chicago Convention. John Cobb Cooper, The Chicago Convention and Outer Space (address before the American Rocket Society Conference, New York, 24 April 1962).

[136] Julian Verplaetse, International Law in Vertical Space, 157 (Rothman, 1960).

[137] Nandasiri Jasentuliyana, International Space Law and the United Nations, 379–382 (Kluwer, 1999).

[138] COPUOS has drafted guidelines and principles, but since the 1979 Moon Agreement, has failed to achieve the consensus necessary to advance a treaty.

[139] Chicago Convention, Article 37. Article 37 of the Chicago Convention may seem to impose no obligation for State to implement standards. However Article 37 and 38 should be read together. This is why ICAO created the safety audit system.

[140] Chicago Convention, Article 12.

CHAPTER 3

SAFETY ISSUES

3.1 Safety issues

Safety standards are typically developed through a consensus process and, in connection with any activity, they are considered to represent the minimal level of risk that society will accept and tolerate. To determine which international standards are needed for space safety, we need first of all to build a common understanding about what are the safety risks posed by the various phases of a commercial space flight, and which ones are of international nature or interest. This section addresses the actual safety challenges that an ICAO-like organization for space will need to address.

3.1.1 Launch site processing and ground safety

On August 22, 2003, at 1330 (local time) a massive explosion destroyed a Brazilian Space Agency VLS-1 rocket as it stood on its launch pad at the Alcantara Launching Centre in northern Brazil. Twenty-one technicians close to the launch pad died when one of the rocket's four first-stage motors ignited accidentally. The rocket had been scheduled to launch in just a few days time and had two satellites onboard when the explosion occurred. The investigation report established that an electrical flaw triggered one of the VLS-1 rocket's four solid fuel motors while it was undergoing final launch preparations. The report said that certain decisions made by managers long before the accident occurred led to a breakdown in safety procedures, routine maintenance, and training. In particular, the investigation committee observed a lack of formal, detailed risk management procedures, especially in the conduct of operations involving preparations for launch.

In the history of humankind, every space adventure, great or small, has begun on the ground. Mission and flight hardware designers who have overlooked this fact have paid a high price, either in loss or damage to the hardware pre-launch, or in mission failure or reduction. Designers may not only risk their flight hardware; they may also risk their own lives, that of their co-workers, and even the safety of the general public by not heeding to calls for safety during ground processing. There are various ground safety issues that apply to all forms of flight hardware from the largest rockets to the smallest spare parts.

A very common issue that the ground processing safety community encounters is lack of recognition of the need for detailed ground safety documentation and rigorous technical safety reviews. Many hardware and mission designers assume that if the hardware is safe to fly, it will also be safe during ground processing. Some also assume that the industrial safety processes used during development and manufacture are sufficient for use at the launch and landing sites.

3.1.2 Flight hardware, ground support equipment, and COTS

At most of the world's launch sites, some form of ground safety review process exist to verify that the flight hardware, its ground support equipment (GSE), and ground operations are in compliance with launch site safety requirements. Nevertheless requirements and risk assessment practices may differ greatly in terms of completeness and adequacy, from country to country, and even from site to site within the same country.

Upon arrival at the launch site, flight hardware and GSE enter into an integrated environment often consisting of ground personnel, the general public, other flight hardware and GSE, and facilities. The safety approval processes should be structured to provide assurances to all interested parties that ground processing hazards are adequately controlled and that the level of residual risk is acceptable.

On many occasions, flight hardware owners focus most of their attention on managing flight hardware risk and not enough attention on GSE risks. Numerous ground processing accidents are directly attributable to the GSE-flight hardware interfaces. For example, the cause (although not the root cause) of the Apollo 13 oxygen tank explosion in flight was the inadvertent damage to the inner part of the tank and its electrical circuitry during ground processing at the Kennedy Space Centre.[1]

The misuse of commercial-off-the-shelf (COTS) equipment is also a major concern. This occurs when a project purchases COTS and then modifies it to meet its specific needs (with the risk of inadvertently removing some safety feature) or when a COTS is used which is not compatible with the flight hardware (e.g. flight ordnance and resistance measurement meters).

An important requirement that many flight hardware developers tend to overlook is the use of written procedures. Written procedures are critical in assuring that operations are run in the manner that the test team intended. The lack of adequate procedures or, even worse, the failure to use written procedures has been listed as the cause of numerous accidents. The timing of the institution of procedures is also important. A procedure that is hurried through development and review and is issued just prior to the operation can be as risky as having no procedure at all. An example of this is the Apollo I fire.[2] The procedure for the "plugs out" test had hundreds of changes made to it and was issued the night before first use. This meant that the test team saw the procedure for the first time during the test. This example illustrates the fact that the use, control, and significance of written procedures cannot be over-emphasized.

Flight hardware providers must take special considerations into account while their hardware is in the ground environment. The same level of vigilance as given to flight operations should be accorded to ground processing safety. Table 3.1 briefly describes some of these issues:

Tab. 3.1: *Sample list of ground processing issues*

Contingency Planning	Unexpected events, whether external to the hardware, such as heavy weather, or internal, such as a propellant leak, must be planned for so that if they occur, prompt corrective action is taken.
Tools	The proper design and use of tools is important to complete pre-launch processing. Accounting for hand tools is critical so they do not become part of the flight hardware.
Chemical and Biological	It is commonly accepted that these commodities, when contained in experiments, present hazards to the flight crews; however, their handling on the ground presents different problems which are often overlooked by developers. This generally applies to sample preparation or return because during these times the commodities are outside their containment. These same issues also apply to the return of trash.
Mechanical Systems (Pressure, Ordnance, Deployable)	All of these systems contain stored energy which must be released in a controlled manner, at the proper time. This is often accomplished through the use of inhibitors. The status of inhibitors on the ground must be carefully tracked in order to not inadvertently activate the system.

[1] http://nssdc.gsfc.nasa.gov/planetary/lunar/ap13acc.html (last accessed: 03 January 2011).
[2] http://history.nasa.gov/Apollo204/ (last accessed: 03 January 2011).

3.2 Launch safety

Space access has become increasingly important to the nations of the world. Upon achieving the status of a space-faring nation however, a key responsibility that devolves upon a State is to establish the technology and processes to protect life and property against the consequences of malfunctioning rockets. Moreover, at most launch ranges the emphasis is on protecting people against injuries resulting from a launch operation. The common practice is to attempt to achieve protection by isolating the hazardous condition from populations at risk. When this is not feasible, risk management may be used to provide an adequately high level of safety. Table 3.2 illustrates how risks can be controlled by a combination of activities.

Identification of high hazard areas may range from simplistic rules of thumb to sophisticated analyses. When simple rules are applied, they commonly specify a hazard radius about a launch point, and planned impact points for stages, connected by some simple corridor. More sophisticated analyses attempt to identify credible rocket malfunctions, model the resulting trajectories, and determine the conditions that will result in debris such as exceeding the structural capacity of the rocket or a flight termination action by a range safety officer.

These analyses typically include failure analyses to identify how a launch vehicle will respond followed by failure response analyses to define the types of malfunction trajectories the vehicles will fly. The vehicle loads are assessed along the malfunction trajectory to determine whether structural limits will be exceeded. Vehicle position and velocity may be compared against abort criteria to assess whether the vehicle should be allowed to continue flight, terminate thrust, or be destroyed. Debris-generating events then become the basis for assessing the flux of debris falling through the atmosphere and the impact probability densities. The debris involved may be screened by size, impact kinetic energy, or other criteria to

Tab. 3.2: *Launch safety risk management*

1	Exclusion of people from high hazard areas
2	Monitoring launch vehicle performance and health
3	Designing and implementing flight termination strategies to limit excursions from the planned trajectory of a malfunctioning vehicle
4	Evaluation of the residual risk after controlling where people may be at risk and the possible excursion of the malfunctioning rocket

assess which fragments pose a threat to unsheltered people, people inside various types of shelters, people on ships, and people in aircraft. The resulting debris impact zones or impact probability isopleths are then commonly used as part of the basis for defining exclusion areas.

Other hazards associated with launch operations are frequently addressed in defining exclusion zones. Explosive hazards (overpressure and fragments thrown by an explosion) are an important component in the launch area. Toxic hazards from the rocket's exhaust products are often an additional consideration in defining exclusion regions. Additional sections of the complex may be restricted to protect against radiation from radars and other support instrumentation. Although full hazard containment is considered to be the preferred protection policy, it is not always possible. The next line of protection after defining exclusion areas is real-time tracking and control of the rockets. Range safety systems are used for this purpose. They include a means of tracking a launch vehicle's position and velocity (tracking system) and a means of terminating the flight of a malfunctioning vehicle (flight termination system).

Flight termination criteria are customarily designed based on the capability of the range safety system to limit the risk from a malfunctioning launch vehicle. Frequently, ranges assume that they can reliably detect a malfunctioning launch vehicle and terminate its flight whenever good quality tracking data is available. This assumption is based on high-reliability designs customarily used for range safety systems. At present, however, there are no international design standards for range safety systems. Moreover, efforts to assure that the design standard does, in fact, achieve the intended reliability levels are rare.

The final tier of protection is risk analysis and risk management. Residual risks from the launch are quantified and assessed to determine if they are acceptable. This step involves an extension of the model outlined above for assessing hazard areas. It is common to perform these protection steps in an iterative manner, using the results of each step to adjust the approach to the others until the desired level of safety is achieved with acceptable impacts on the proposed launch. The current practice is to assess risks for each launch and to approve the launch only when risk levels are acceptable. Unlike most other activities, annual risk levels are addressed by exception.

A proper risk analysis addresses the credible risks from all launch-related hazards. These may include inert debris, fire-brands, overpressure from exploding fragments, and toxic substances generated by normal combustion as well as toxic releases from malfunctions. In practice, for many launches the contribution from one or more hazards can be demonstrated to be negligible. When these contributions are not demonstrably negligible, appropriate risk estimates are desired.

Current practice often fails to properly address toxic hazards, explosive hazards, and hazards from firebrands. It is also inconsistent in addressing initiating events.

Some ranges attempt to quantify risks from all debris-generating events, including malfunctions and launch vehicles; others focus on planned events, such as jettisoning spent stages.

When assessing launch risks it is important to account for all exposed populations: people on land, people in ships, and people in aircraft. Proper consideration must be given to the effect of sheltering on the risks. It is often assumed that neglecting sheltering will overstate the risk. When sheltering is adequate to preclude fragment penetration, this assumption is valid. When fragments are capable of penetrating a structure, debris from the structure increases the threat to its occupants.

Additional consideration must be given to the relationship of population groups to the launch. People directly involved in supporting launch operations may be expected to tolerate higher risk levels than members of the general public who are not involved. Typically, launch support personnel are confined to a region in the vicinity of the launch point within the territorial domain of the launching nation. As launch vehicles proceed downrange, they typically leave the territorial domain of the launching nation and begin to over-fly international waters and the territory of other nations.

> Current practice addresses jurisdictional differences minimally. People at risk from a launch are typically provided no additional consideration when they are located in international waters or outside of the domain of the launching country.

Tolerable risks for a launch are commonly expressed in terms of a collective or societal risk level and risk to the maximally exposed individual (individual risk). Collective risk is commonly expressed as the number of individuals statistically expected to be exposed to a specified injury level. Individual risk is commonly expressed as the probability that the maximally exposed individual will suffer the specified injury level. The two most commonly used levels of injury are fatality and serious injury. Serious injury is often correlated with level 3 or greater of the Abbreviated Injury Scale (AIS). When it is difficult to quantify risk directly, impact probability for specified classes of debris is often used as a proxy measure. Thus, for example, it is customary to protect people on ships or people on airplanes by creating exclusion zones based on impact probabilities.

Historically, many ranges computed impact probability for these decisions based upon the most comprehensive debris lists they could obtain. Alternatively, many have based their decisions on protecting against "hazardous debris". The definition of hazardous debris for this purpose has varied between ranges and over the passage of time. Recent efforts to standardize the definition of "hazardous debris" include the U.S. Range Commanders Counsel Publication RCC 321-99 and its most recent update, RCC 321-07.

When an exclusion area is defined, each nation has its own procedures for communicating the boundaries of the area. On land, this is commonly through sign postings and guards. Formal notices are frequently used to communicate with operators of ships and aircraft. Moreover, the degree of compliance varies with location and time. When the exclusion area is near the launch complex, ranges frequently employ some form of surveillance to determine whether any vessels have intruded into the hazardous area. When intruders can be identified, the ranges may request them to depart, passively wait for their departure, or proceed with the launch based on the decision that the risk to the vessel is sufficiently small.

Outside of the immediate launch area, surveillance becomes more difficult and more costly. Consequently, most ranges use surveillance very selectively outside of the immediate launch area, typically restricting surveillance to planned impact areas for spent stages and other planned jettisons. As a result, publishing exclusion areas at these distances is much less effective. More efficient tools for surveying these remote locations and communicating with intruders would enhance the effectiveness of protecting ships and aircraft in these areas.

Controlling risks to seafaring vessels from space launch testing activities is most successful when mariners are notified about hazard areas and when the responsible launching agency surveys the potentially affected areas to detect intruders and to warn them to leave the area. Following a mishap, communication with these vessels to proceed at maximum speed in a prescribed direction to minimize impact probability is essential to control undue risks. Currently, costs and technology limit surveillance and communication to locations near land.

> At present each nation with space access capability determines *independently* what risk limits to other nations are acceptable.

Management of airspace must consider aircraft traffic. At present, there are limited capabilities for addressing these issues. The FAA has begun an initiative to address these concerns for U.S. operations. It should be noted that the current practice is for each range to manage risks on a mission-by-mission basis through Launch Collision Avoidance (LCOLA) processes. Minimal attention is paid to annual risks generated by the range's launch operations. There is no agency – national or international – that monitors and controls risk posed to overflown populations. A city may be placed at risk by launches from multiple launch sites without the performance by involved launching nations of any coordinated assessment to assure that the risk levels are acceptable.

Citizens of all nations should be equally protected from the risk posed from overflying by launch vehicles and returning spacecraft. The common practice is to make these determinations on a launch-by-launch basis with no consideration of previous or planned future launches. As a result, the likelihood increases that a nation that is subjected to overflight will endure significant annual risks from (1) a single launch facility, (2) a single nation's activity, or (3) all nations' launch activities.

Finally, there is the health risk related to the dropping of rocket stages and ascent failures. During normal launches, stages separate sequentially and fall down to Earth. Most launch trajectories and spaceport locations are chosen to ensure that the impact areas are outside populated areas and are mainly contiguous to the oceans. Nevertheless there are inland spaceport locations and land overflying trajectories which lead to stages dropping to ground in sparsely inhabited areas with ensuing soil contamination. Approximately 9% of the propellant from a launch stage remains in the tank once it is dropped.[3] The penetration of contaminants depends on the nature and properties of the soil and can lead to the contamination of groundwater as well as surface water. For example, hydrazine (UDMH) is often used in hypergolic rocket fuels as a bipropellant in combination with the oxidizer nitrogen tetroxide and less frequently with IRFNA (red fuming nitric acid) or liquid oxygen. UDMH is a toxic carcinogen and can explode in the presence of oxidisers. It can also be absorbed through the skin. A tablespoon of hydrazine in a swimming pool would kill anyone who drank the water. In a study conducted by Vector, the Russian State Research Center of Virology and Biotechnology in Novosibirsk, health records from 1998 to 2000 of about 1000 children in two areas in southern Siberia polluted due to launches from Baikonur in Kazakhstan were examined, comparing them with 330 records from a nearby unpolluted control area. Grouping all cases of disease together, the research team concluded that children from the worst affected area were up to twice as likely to require medical attention for diseases such as endocrine and blood disorders during the three years studied and needed to be treated for twice as long.[4] Contamination can be far worse and massive in case of launch failure. In September 2007, the explosion of a Russian Proton M rocket contaminated a vast swath of agricultural land in Kazakhstan with 200 metric tons of toxic fuel.

[3] Eco Space Final Report, SSP 2010, International Space University, Strasbourg, France (2010).
[4] Nature, "Study links sickness to Russian launch site", 433, 95 (13 January 2005).

3.3 Suborbital safety

A suborbital flight is defined as a flight up to a very high altitude beyond 100 km above sea level but in which the vehicle involved does not go into orbit (i.e., does not attain an orbital speed exceeding 11.2 km/s). A "suborbital trajectory" is defined under U.S. law as: "The intentional flight path of a launch vehicle, re-entry vehicle, or any portion thereof, whose vacuum instantaneous impact point does not leave the surface of the Earth".[5] Unmanned suborbital flights have been common since the very beginning of the space age. Sounding rockets covering a wide range of apogees even well above the altitude of the Shuttle and ISS orbits have been routinely launched. Figure 3.1 illustrates the variety of suborbital rockets used by the European Space Agency for microgravity experiments and their maximum altitude compared to orbital vehicles.

Nowadays, suborbital human spaceflight is gaining popularity as demonstrated by the increased interest in space tourism. Still in its nascent phase, the space tourism industry proposes new vehicle configurations and related safety risks substantially similar to those of the previous era. In May 1961, Alan Sheppard reached an altitude of 187 km on board the first Mercury man-rated rocket (Mercury Redstone 3, a rocket with a capsule on top). Just over two years later,

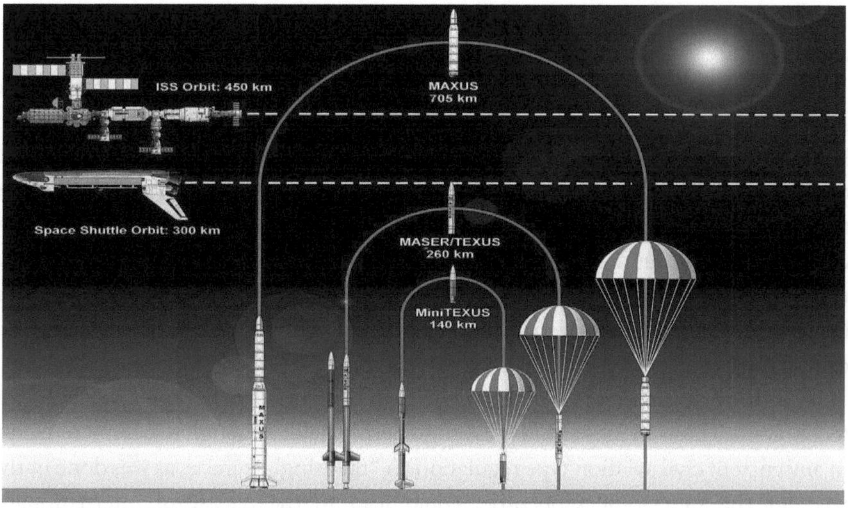

Fig. 3.1. *Suborbital rockets altitudes (Credits: © ESA/G. Dechiara).*

(1)

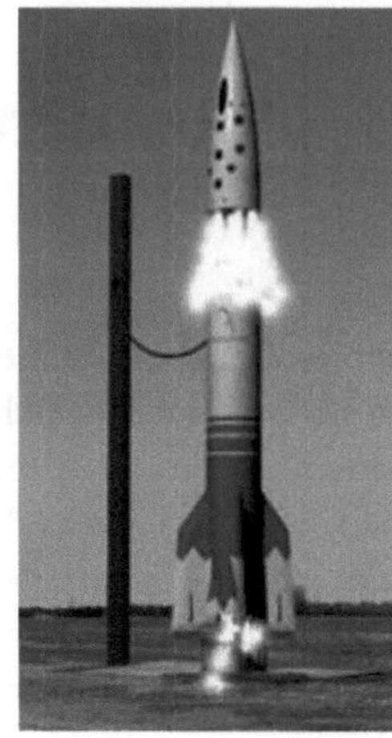

(2)

(1) Configuration Aircraft: *Ascender*

(2) Configuration Launcher/capsule: *Canadian Arrow* in November 2010 The President of Canadian Arrow stated that the company is now likely not going to fly Canadian Arrow Rocket as space tourism vehicle.

Fig. 3.2. *Suborbital vehicles configurations (Credits: © Bristol Spaceplanes Ltd and © Canadian Arrow).*

NASA test pilot Joseph Walker reached an altitude of 108 km in an X-15 aircraft, and returned to the runway from which he took off (attached to a B-52 mother ship). The X-15 was an airplane designed to explore the lower end of the hypersonic flight regime, and it had been supplemented with a reaction control system (RCS) to manoeuvre at altitudes where aerodynamic forces were negligible. The Mercury Redstone, on the other hand, was a spacecraft capsule on top of a man-rated launcher. The commercial manned space vehicles currently in development still basically follow one of the two basic configurations described above. Figure 3.2 illustrates the two kinds of suborbital vehicles.

The aircraft-type configuration has limits that constrain it to operate around the Karman Line. The development of space-planes equally capable of sustained flight in the atmosphere and above is still far away in the future because of technological difficulties in developing the propulsion system. It should be noted that the two configurations drive very different safety requirements. Safety requirements (developed by the launch site, under the jurisdiction of its national government) for the launcher/capsule configuration have been in place for more than 40 years and have been successfully proven, mainly during the conduct of (more challenging) orbital flights.

The safety requirements for the aircraft-type configuration have a well established technological basis in the aeronautical engineering field, although this is not reflected in any current civil aviation type regulation. A "tailoring" exercise, as was done in the past for the supersonic Concorde, could lead to the release of internationally recognized flightworthiness standards in a relatively short time frame.

The capsule on top of a launcher configuration has proven to be expensive but safe, while the airplane-type configuration is less expensive but also less safe. The experimental aircraft X-15 flew another 199 times before cancellation of the programme in 1968. The X-15 suffered four major accidents, one of which involved a human casualty (Maj. Michael Adams). The comparable fatality risk for this configuration is therefore at best in the order of 1 in 200 flights.

[5] Commercial Space Launch Act, 49 USC § 70102 (20) (2004). Available online: http://uscode.house.gov/download/pls/49C701.txt (last accessed: 03 January 2011).

3.4 Orbital safety issues

The principal safety issues related to orbital human spaceflight are: protection from dangers in the space environment (debris and ionizing radiation); the provision of escape capabilities during the launch and on-orbit phases; and the prevention of collision risk. Collision risk may be divided into (1) the risk of collision during proximity operations (i.e., rendezvous and docking), which is part of the more general safety issue of systems interoperability and (2) the risk of collision with general air and space traffic.

3.4.1 Orbital debris

Space is not an empty vacuum but contains both natural debris (i.e. micrometeoroids, interplanetary dust) and human-made space junk. Humans generally

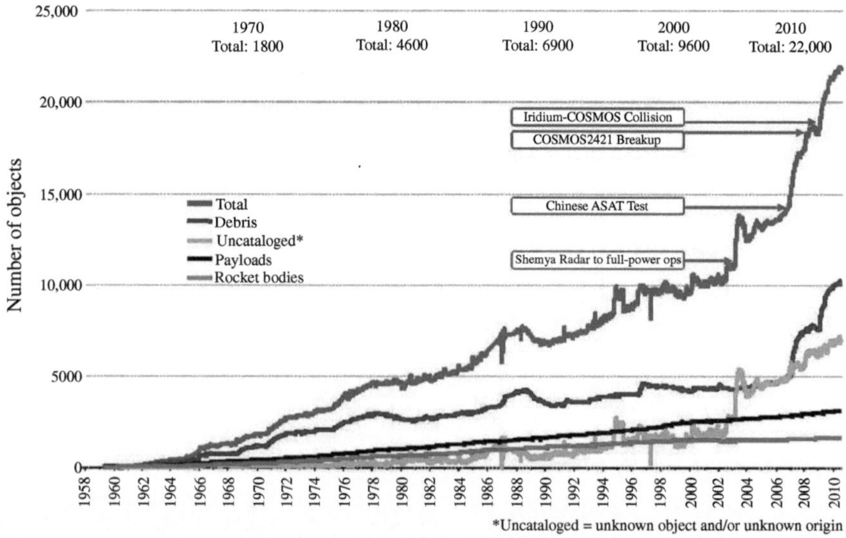

Fig. 3.3. *Satellite catalogue growth.*[6]

have no involvement in the creation of natural debris, thus here we will concentrate exclusively on human-made debris. Orbital debris generally refers to any human-made material on orbit which is no longer serving any useful function. There are many sources of debris. One source is discarded hardware such as upper stages of launch vehicles or satellites which have been abandoned at the end of their useful lives. Another source is spacecraft items released in the course of mission operations. Typically, these items include launch vehicle fairings, separation bolts, clamp bands, adapter shrouds, lens caps, momentum flywheels, and auxiliary motors. Various shapes and sizes of debris are also produced as a result of the degradation of hardware due to atomic oxygen, solar heating, and solar radiation, and also from combustion of solid rocket motors. Examples of such products are paint flakes, aluminium oxide exhaust particles, and motor-liner residuals.

Fifty years of spaceflight have cluttered the space around the Earth with an enormous quantity of human-made debris. Scientists assume that there are approximately 500,000 objects in orbit whose sizes are above 1 centimetre. Currently, about 21,000 of such objects (i.e.: 10 cm in diameter or larger) are being tracked by the U.S. Space Surveillance Network (including about 800 objects representing functional satellites). Only the largest pieces of debris in orbit can be regularly tracked, mainly by using optical sensors. In the geosynchronous orbit, the minimum size that can be tracked is 30 cm, while in low Earth orbits it is about 10 cm. Among the tracked pieces of debris, there are about 200 satellites abandoned in geosynchronous orbits occupying or drifting through valuable orbital positions and posing a collision hazard for functional spacecraft. The survival time of the debris can be very long. Objects in 1000 km orbits can exist for hundreds of years. At 1500 km, the lifetime can go up to thousands of years. Objects in geosynchronous orbit can presumably survive for one million years.

The amount of debris on orbit in the future will depend upon whether the creation or removal rate dominates. Currently, the only mechanism for removal of debris is orbital decay through atmospheric drag, which ultimately leads to re-entry. This mechanism is only effective in a restricted range of low Earth orbits. At higher orbits, it takes hundreds to thousands of years for objects to re-enter the Earth's atmosphere. Consequently, there is no effective removal mechanism. Historically, the creation rate of debris has outpaced the removal rate, leading to a net growth in the debris population in low Earth orbit at an average rate of approximately 5% per year. A major contributor to the current debris population has been fragment generation via explosions. As the debris mitigation measure of passivation (e.g., depletion of residual fuel) comes to be practiced more commonly, it is expected that explosions will decrease in frequency. It may take a few decades for the practice to become implemented widely enough to reduce the explosion rate, which currently stands at about four per year.

Several environment projection studies conducted in recent years indicate that, with various assumed future launch rates, the debris populations at some altitudes in low Earth orbit will become unstable. Collisions will take over as the dominant debris generation mechanism, and the debris generated will feed back into the environment and induce more collisions. According to Liou and Johnson,[7] the most active orbital region is between the altitudes of 900 and 1000 km and, even without any new launches, this region is highly unstable. It is projected that the debris population (i.e.: objects 10 cm and larger) in this "red zone" will approximately triple in the next 200 years, leading to an increase in collision probability among objects in this region by a factor of ten. In reality, the future debris environment is likely to be worse than as suggested by Liou and Johnson, as satellites continue to be launched into space. Their paper concludes that to better limit the growth of future debris populations, active removal of objects from space needs to be considered.

In 2009, the Tactical Technological Office at DARPA initiated a study called "Catcher's Mitt" to model the debris problem and its future growth, determine where the greatest problem will be for assets, and then explore technically and economically feasible solutions for debris removal. As part of the study, DARPA issued in September 2009 a Request for Information (RFI) to identify promising technical approaches and system concepts for cost effective and innovative removal of orbital debris.[8] The RFI stated that the debris population of interest included small (1–10 mm), medium (1–10 cm) and large (derelict spacecraft/expended rocket bodies) sized debris in low Earth orbit (LEO), as well as large sized debris in Geosynchronous Earth Orbit (GEO). One of the responses to this RFI advocated the use of high energy laser for space debris removal.[9]

3.4.2 Collision risk with orbital debris

Orbital debris generally moves at very high speeds relative to operational satellites. In low Earth orbit (i.e., altitudes lower than 2000 km), the average relative impact velocity is 10 km/s (36,000 km/hr). In the geosynchronous orbit, the relative velocity is lower, approximately 2 km/s, because most objects move in an eastward direction. At these hyper velocities, pieces of debris have a tremendous amount of kinetic energy. A 1 kg object, moving at a speed of 10 km/s has the same amount of kinetic energy as a fully loaded truck, weighing 35,000 kg, has at 190 km/hr. A 1 cm sized aluminium sphere at orbital speed has the energy equivalent of an exploding hand grenade. A 10 cm fragment in geosynchronous orbit has roughly the same damage potential as a 1 cm fragment in low Earth orbit.

Pieces or particles of debris smaller than 1 mm in size do not generally pose a hazard to spacecraft functionality. Debris fragments from 1 mm to 1 cm in size may or may not penetrate a spacecraft, depending on the material composition of the debris and whether or not shielding is used by the spacecraft. Penetration through a critical component, such as the flight computer or propellant tank, can result in loss of the spacecraft. NASA considers pieces of debris 3 mm in size and above as potentially lethal to the Space Shuttle and the International Space Station. Debris fragments between 1 and 10 cm in size will penetrate and damage most spacecraft. If the spacecraft is impacted, satellite function will be terminated and, at the same time, a significant amount of small debris will be created.

If a 10 cm debris fragment weighing 1 kg collides with a typical 1200 kg spacecraft, over one million fragments ranging in size from about 1 mm and larger could be created. Such collisions result in the formation of a debris cloud which poses a magnified impact risk to any other spacecraft in the orbital vicinity (e.g., other members of a constellation of satellites).

Certain regions of the debris cloud are constricted to one or two dimensions. Such constrictions do not move with the debris cloud around its orbit. They remain fixed in inertial space while the debris cloud repeatedly circulates through them. In many satellite constellations, there are multiple satellites in each orbital ring. If one of these satellites breaks up, the remaining satellites in the ring will all repeatedly fly through the constrictions. If many fragments are produced by the breakup, the risk of damaging another satellite in the ring may be significant. If satellites from two orbital rings collide, two debris clouds will be formed with one in each ring. The constrictions of each cloud will then pose a hazard to the remaining satellites in both rings.

In February 2009, a non-operational Russian satellite, Cosmos 2251, collided with Iridium 33, a U.S. commercial telecommunication satellite, over Siberia at an altitude of 790 km. This collision, the first of its kind, was the worst space debris event since China intentionally destroyed one of its aging weather satellites during an ASAT test in 2007. The Iridium satellite that was lost in the collision was part of a constellation of 66 low Earth orbiting satellites providing mobile voice and data communications services globally. As expected, the risk of collision of other Iridium satellites in the same plane dramatically increased with daily announcements of possible collisions (called conjunctions) with the debris from Iridium 33.

3.4.3 Collision risk to human spaceflight

Orbital debris collision is the primary source of risk for the International Space Station, and accounts for 11 of the 20 potential problems most likely to cause the loss of a Shuttle and crew.

The 2003 Shuttle risk assessment performed after the Columbia accident was the first one that incorporated the threat posed by orbital debris. It determined that the likelihood of orbital debris bringing down the Shuttle is far greater than that of the widely feared failures of main engines, solid rocket boosters, or thermal protection. Orbital debris colliding with different spots of the wing flaps is the most likely catastrophic failure. Damage could render a wing flap (elevon), unable to steer and slow the Shuttle as it plummets through the atmosphere and would cause its inevitable loss. Following the Shuttle STS 114 mission in the summer of 2005 (i.e., the return to flight mission after the Columbia accident), NASA inspection teams catalogued 41 orbital debris impact locations on the Shuttle vehicle (Orbiter). There were fourteen impacts on the windows, and two windows had to be replaced (as had happened several times in the past). The largest impact, featuring a 6.6 mm × 5.8 mm crater, was caused by a particle with an

Tab. 3.3: *Controlling orbital debris risk*

Passivation	To prevent explosions, satellite components that store energy can be passivated at the end of their useful life. For example, propellant in upper stages and satellites can be eliminated by either venting or burning to depletion. Batteries can be also designed to reduce risk of explosion.
Collision avoidance manoeuvres	Spacecraft manoeuvres can also mitigate orbital debris risk of collision. The Space Shuttle and the International Space Station have been manoeuvred on several occasions to avoid collisions with orbital debris. Also, in the case of satellite constellations, because a potential collision will lead to the creation of a debris cloud that may result in damage to other members of the constellation, collision avoidance manoeuvres may be necessary.
Shielding	Protection by shielding can be used to minimize the risk, but, of course, there are mass, and therefore performance, penalties that make such solutions feasible only under exceptional circumstances. This is the case of the International Space Station (ISS). All together there are 100 different shields protecting the ISS. The ISS is the most heavily shielded spacecraft ever flown. Critical components such as habitable compartments and high-pressure tanks will be able to withstand the impact of debris as large as 1 cm in diameter.
End-of-life disposal	To prevent debris accumulation in preferred mission orbits due to collisions, satellites and other objects must be removed from the mission orbit at the end of life. Guidelines currently adopted by a number of government organizations recommend that an object should not remain in its mission orbit for more than 25 years. Satellites, upper stages, and deployed objects removed to sufficiently low altitudes in low Earth orbit are subjected to a residual atmospheric drag which would cause the object's orbit to decay naturally and result in re-entry within 25 years.

estimated diameter of 0.22 mm. The impact was among the largest ever recorded on a crew module window.[10]

Orbital debris risk is best controlled by limiting the creation of debris through a number of measures which usually increase development and operating costs. Table 3.3 lists different methods for controlling orbital debris risk.

At altitudes above 2000 km, it is not economically feasible in most cases to force re-entry of debris within 25 years. At this time, it is generally recommended to place vehicles in disposal (or "graveyard") orbits. Many spacecraft that were previously operating in the geosynchronous orbit have already been boosted into a higher disposal orbit at the end of their mission life. Propellant must be reserved to perform the disposal manoeuvres. Hence, the cost to satellite operators is reduced mission life, and to launcher operators, it is reduced performance of upper stages. Estimates of the amount of "wasted" lifetime for geosynchronous satellites vary between six months and two years. For example, it has been calculated that if a typical commercial communication satellite that has 24 Ku-band and 24 C-band transponders with bandwidths of 36 MHz has to be boosted into a higher disposal orbit at the end of its mission life, this manoeuvre would cause the satellite operator an average loss (in terms of how much longer the satellite could have continued commercial operations) of one year's profit, equivalent to at least one hundred million U.S. dollars.[11]

Finally, to mitigate the debris risk, an international treaty is needed to ban anti-satellite (ASAT) tests. On January 11, 2007, a 958 kg Chinese weather satellite Fengyun-1C was destroyed by a medium-range missile in a Chinese ASAT test.[12] This test created fragments with apogees reaching up to 3500 km and perigees below 200 km altitude, which will stay on-orbit for hundreds of years and which represent the second worse event ever in terms of orbital debris contamination of the orbital environment. Because of the test, the risk to the International Space Station from fragment sizes bigger than 1 cm increased by 59%. At this size, debris cannot be spotted from the ground and, yet, it is beyond the capability of the shielding used on the ISS.

3.4.4 Orbital debris ground risk

As previously mentioned, non-functional satellites, spent launch vehicle upper stages, and other orbital debris do not remain in their orbits indefinitely, but gradually return to Earth. However, in low Earth orbits, the re-entry may take years or tens of years to occur, and at higher altitudes, based on periodic fluctuation in the atmospheric density, it may take hundreds or even thousands of years.

As non-functional satellites, spent launch vehicle stages and other pieces of debris lose altitude. They enter denser regions of the atmosphere where friction with atmospheric gases at high velocity generates a tremendous amount of heat. As a result, a major portion of the hardware (between 60 and 90%) will burn up. However, some components and parts can and do survive the re-entry heating.

Component survival will occur if the melting temperature of the component is sufficiently high or if its shape enables it to lose heat fast enough to keep the temperature below the melting point. During re-entry, the object is decelerating quickly and the structural loads can exceed 10 g (10 times the acceleration of gravity). These loads, combined with the high temperatures, cause fragmentation to occur. When the resulting objects lose enough speed, the heating rate is reduced, the temperature decreases, and the objects begin to cool. By this time, the objects have fallen to even denser regions of the atmosphere and fall virtually straight down from the sky. They impact the ground at relatively low speeds, but still represent a potential hazard to people and property on the ground. They also represent a serious risk to maritime and air traffic.

> On average, one non-functional satellite, launch vehicle orbital stage, or other piece of catalogued debris re-enters Earth's atmosphere every day.

It is very difficult to predict where debris from a randomly re-entering satellite will hit the surface of the Earth. Over the last 50 years, more than 1400 metric tons of materials are believed to have survived re-entry with no reported casualties. The largest object to re-enter was the Russian Mir Space Station, which weighed 120,000 kg. More than fifty pieces of debris were recovered and documented over the years. The items shown in Figure 3.4 represent some examples of re-entered materials ranging in weight between 30 and 250 kg. These in particular were all identified as upper stages of Delta 2 rockets used to launch GPS satellites and they re-entered between 1997 and 2001 in different places on the globe. In 2004 and 2005, the same type of Delta 2 titanium motor casings reached the ground in Argentina and Thailand respectively. Another seven Delta 2 titanium motor casings re-entered in the period 2001–2005 and probably fell into the ocean.

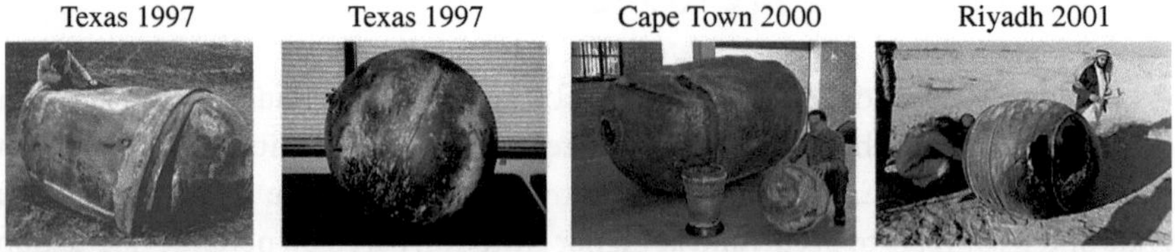

Fig. 3.4. *Fallen orbital debris.*

In general, components made of aluminium and similar materials with low melting temperatures do not survive re-entry while pieces or components made of materials with high melting temperatures, such as stainless steel, titanium, and glass often do survive. Pieces that survive re-entry tend to be large and, in some cases, heavy.

Currently, it is considered that although such catastrophic risk, including the possibility of injury to large numbers of people, is very low, it would still not be acceptable to the general public, and such an accident would expose the launching nation to legal liability to pay compensation. For such reasons, mission requirements in a number of countries prescribe that the risk of any personal casualty due to a single re-entry event must be less than 1 in 10,000 re-entries.

On 21 February 2008, an uncontrolled re-entering satellite was shot down on officially stated grounds of public safety. The satellite was destroyed at an altitude of 247 km in space by a three-stage Standard Missile-3, which was a modified version of an existing missile adapted to intercept ballistic missiles in flight. The intercept was planned so as to create only short-lived space debris. The decision was made at the U.S. presidential level. The malfunctioning spacecraft, a U.S. spy satellite (USA 193), carried 450 kg of highly toxic frozen hydrazine fuel in its titanium fuel tank. Similar tanks are known to have survived re-entry. However, following piping rupture and metal softening due to re-entry heat, the normally unfrozen fuel had completely leaked out and had been dispersed high in the atmosphere. Similar titanium tanks of the ill-fated Shuttle Columbia survived re-entry but when found were practically empty. In any case, it was expected that about 50% of the U.S. spy satellite's mass of 2270 kg would survive re-entry, thus adding to public risk on ground.

> The total amount of radioactive fuel in low Earth orbit is approximately 1,000 kg resulting from planned separation at the end of the mission of the nuclear reactors and their injection into disposal orbits.

Currently, there are 32 defunct nuclear reactors circling the Earth as well as thirteen reactor fuel cores and at least eight radio-thermal generators (RTGs). RTGs had been used six times in space missions in low Earth orbits up to 1972, and twice in the geosynchronous orbit up to 1976. Since 1969, another fourteen reactors have been used on lunar and interplanetary missions. The total mass of RTG nuclear fuel in Earth orbit today is in the order of 150 kg. Another form of nuclear power source used in space activities is a nuclear reactor. Most of these reactors were deployed on Soviet radar reconnaissance satellites (RORSATs) launched between 1965 and 1988.

Among the space nuclear accidents on record, (i.e., unwanted/unplanned release of radioactive material), two involved orbital debris, and a third was a close call. In 1978, the RORSAT COSMOS 954 failed to separate its nuclear reactor core and to boost it into a disposal orbit as planned. The reactor remained onboard the satellite in an orbit that decayed until it re-entered the Earth's atmosphere. The satellite crashed near the Great Slave Lake in Canada's Northwest Territories, spreading its radioactive fuel over an area of about 124,000 km^2. Recovery teams swept the area by foot for months. Ultimately, they were able only to recover twelve large pieces which comprised a mere 1% of the estimated quantity of radioactive fuel on board. These pieces emitted radioactivity of up to 1.1 Sieverts per hour. (Usually a nuclear emergency is declared on ground at 500 micro-Sieverts per hour.)

A few years later, in 1982, another RORSAT, COSMOS 1402, failed to boost the nuclear reactor core into a storage orbit. The ground controller managed to separate the core from the reactor itself to make it more likely that it would burn up in the atmosphere before reaching the ground. The reactor was the last piece of the satellite to return to Earth in February 1983 when its core fell into the South Atlantic Ocean.

Lastly, in April 1988, yet another Russian spacecraft, COSMOS 1900, failed again to separate and boost the reactor core into a storage orbit. However, later on, the redundant system succeeded in separating and boosting the nuclear core into a storage orbit, although lower than that originally planned.

[6] Source: Joint Space Operations Center, cited in National Security Space Strategy-Unclassified Summary, January 2011, page 1.

[7] J.-C. Liou and N.L. Johnson, "Risk in Space from Orbiting Debris", Science, Vol. 311, 20 (January 2006).

[8] https://www.fbo.gov/index?s=opportunity&mode=form&id=a55fd6e5721284ee7df2068d2b300b5f &tab=core&_cview=0 (last accessed: 03 January 2011).

[9] C.P.J. Barty, J.A. Caird, A.E. Erlandson, R. Beach, A.M. Rubenchik, "High Energy Laser for Space Debris Removal" (October 31, 2009). Available online: https://e-reports-ext.llnl.gov/pdf/381096.pdf (last accessed: 03 January 2011).

[10] NASA, Orbital Debris Quarterly News, Volume 10, Issue 3, 2 (July 2006). Online: http:// orbitaldebris.jsc.nasa.gov/newsletter/pdfs/ODQNv10i3.pdf (last accessed: 03 January 2011).

[11] K. K. Galabova and O. L. Weck, "Economic case for the retirement of geosynchronous communication satellites via space tugs", Acta Astronautica, Vol. 58, 485–498 (2006).

[12] C. Pardini and L. Anselmo, "Assessment of the consequences of the Fengyun-1 C breakup in low Earth orbit", Advances in Space Research, Vol. 44, Issue 5, 545–557 (2009).

3.5 Returning vehicles risk

The disintegration during re-entry of the Shuttle Columbia on February 1, 2003 was a watershed moment in the history of launch and re-entry safety analysis. It highlighted the need to select vehicle re-entry trajectories which minimize the risk to ground populations, and the need to take measures to keep air traffic away from falling debris if a re-entry accident occurs. The Columbia accident initiated a chain of events that demonstrated the need for a deliberate, integrated, and, eventually, international approach to public safety during launch and re-entry operations. This is especially true for the management of air traffic and space operations.

Table 3.4 lists some re-entry safety questions. Shortly after the breakup of Columbia over a relatively sparsely populated area of Texas,[13] dramatic images of the debris from the breakup of the orbiter were seen around the globe: an intact spherical tank in a school parking lot, an obliterated office rooftop, mangled metal along roadsides, and charred chunks of material in fields. The NASA Administrator testified before the U.S. Senate that it was "amazing that there were no other collateral damage" (i.e., that no members of the public were hurt).[14] Some people wondered if it was a "miracle" that no one on the ground had been hurt, and raised some important questions about public safety during re-entry.

3.5.1 Risk to people on the ground

The Columbia Accident Investigation Board (CAIB) raised and answered many questions relevant to public safety during launch and in particular re-entry. Given

Tab. 3.4: *Key re entry safety questions*

	What were the chances that the general public could have been hurt by the break-up of *Columbia*?
	How safe are Space Shuttle flights compared to the overflight of conventional aircraft?
	How much public risk from space flight is acceptable?
	Who is responsible for public safety during space flight operations?
	What should be done to protect aircraft from potential impacts by debris from space operations?

the available data on the debris recovered and the population characteristics in the vicinity, a CAIB study found that the absence of ground casualties was, in fact, the statistically expected result. Specifically, based on census data and modelling methods consistent with U.S. standards[15] and requirements set by other U.S. agencies (e.g., the USAF in the Air Force Space Command Manual[16] and by the Federal Aviation Administration (FAA) in the Federal Register,[17] the study found that "the lack of casualties was the expected event, but there was a reasonable probability (less than 0.5 but greater than 0.05)[18] that casualties could have occurred".[19] However, a similar event over a densely populated area such as Houston would almost certainly have produced multiple casualties among the public on the ground.[20]

At the time of the Columbia accident, NASA had no formal policy regarding public risk during Shuttle re-entry. NASA's Associate Administrator for Safety and Mission Assurance was quoted in a newspaper article on this subject as saying, "And so what our assessment says is that if this thing is safe enough to fly human beings in for an entry and a landing, then we feel that that's adequate safety for the public that's underneath the flight path."[21] The CAIB disagreed with that approach, and made the following specific recommendations for public safety (Table 3.5).

CAIB observation 10.1-1 led to the development of NASA's new safety policy (NPR8715.5). The NASA public safety policy embraces many of the risk measures and thresholds already in use by other U.S. agencies, such as individual and collective risk limits in terms of casualties.[22] However, NASA's public safety policy also put forward innovative criteria for risk budgets governing distinct phases of flight which have gained broad acceptance.[23] Therefore, the Columbia accident led to greater consensus and innovation in the management of risk to people on the ground from launch and re-entry operations.

Tab. 3.5: *CAIB recommendations (selected)*

O10.1-1	NASA should develop and implement a public risk acceptability policy for launch and re-entry of space vehicles and unmanned aircraft
O10.1-2	NASA should develop and implement a plan to mitigate the risk that Shuttle flights pose to the general public
O10.1-3	NASA should study the debris recovered from *Columbia* to facilitate realistic estimates of the risk to the public during Orbiter re-entry

3.5.2 Risk to people in aircraft

The Columbia accident also promoted the development of improved methods and standards for aircraft safety during launch and re-entry. Following the release of the final report of the CAIB, the FAA funded a more detailed aircraft risk analysis that used the actual records of aircraft activity at the time of the accident. That study found that the probability of an impact between Columbia debris and commercial aircraft in the vicinity was at least one in a thousand, and the chance of an impact with a general aviation aircraft was at least one in a hundred.[24] The analysis used the current models which assume that any impact anywhere on a commercial transport with debris of mass above 300 g produces a catastrophic accident: all people on board are killed. Current best practices are captured in RCC 321-07 "Common Risk Criteria for the National Ranges", which provides a vulnerability model for the commercial transport class. (In 2008, the FAA and USAF sponsored the development of vulnerability models for transoceanic business jets based on the same methods.)

After FAA executives were briefed about the potential for aircraft impacts during the *Columbia* accident, and the challenges presented by integration of innovative vehicles such as the suborbital SpaceShipTwo into the National Airspace System (NAS), the FAA investigated the potential for new decision support tools (DSTs) to better manage the interface of space and air traffic.[25]

The FAA's Space and Air Traffic Management System (SATMS) needs to accommodate any future growth in all kinds of space operations (civil, commercial, and military) in the National Air Space (NAS). In this context, space operations refer to any phase of flight of a space vehicle that may threaten people on the surface of the Earth or in the Earth's atmosphere. The FAA's new DSTs will help controllers manage a diverse mix of aircraft and space vehicles operating in shared airspace.[26] The FAA Office of Commercial Space Transportation (AST) is leading the development of a SATMS DST that is both an off-line planning tool and a real-time operations tool. The plan is for controllers to use the DST to strategically plan and coordinate space operations in the NAS. Also, the DST will be used as a real-time tactical tool in the event of a catastrophic event like the Columbia accident to identify how to re-direct aircraft around a space vehicle debris hazard area. Effective implementation of such a tool may require:

- Standards on the level of protection from spacecraft hazards required by aircraft, which have already been proposed;[27]
- A new generation of communications, navigation, and surveillance services infrastructure (communications network, space vehicle tracking capabilities, etc.).[28]

The primary function of a SATMS DST will be to facilitate optimal use of the NAS by mitigating the risk to aircraft from space operations. The current procedure for ensuring that aircraft are at safe distances during a launch involves restricting all air traffic from flying in a very large region of Special Use Airspace (SUA) and/or Altitude Reservations (ALTRVs). During a launch, aircraft that would normally fly through this airspace must take an alternate route to their destination. The SATMS DST would minimize the impact of the SUA/ALTRV on the NAS by optimizing affected routes around the restricted airspace.[29] For example, the DST would treat the potential debris hazard like an area of severe weather, and would provide conflict advisories and recommend routes to safely and efficiently direct aircraft around the hazard area. In addition, this tool could minimize the footprint of space operations on the NAS by allowing airspace to remain unrestricted unless an accident occurs.

The size and duration of a launch area SUA/ALTRV may be based in part on the results of currently available spacecraft debris impact dispersion and risk models. The proposed SATMS DST would integrate a debris model, such as the one applied to the Columbia accident,[30] with an aircraft trajectory model similar to those used in current air traffic management (ATM) tools. As a result, the planned DST will not necessitate the development of new technologies. Rather, it will integrate existing ATM and range safety technologies. However, a primary challenge in the development of any space and air traffic management system will be to establish how safe is safe enough for the aircraft flying in the vicinity of a launch or re-entry. Other challenges will be (1) the coordination of organizations responsible for the control of air and space operations across the globe and (2) determination of how the information is best presented to controllers of aircraft and space vehicle movements.

[13] The disintegration of Columbia occurred over an area with an average of about 70 inhabitants per square mile.

[14] Columbia Accident Investigation Board (CAIB), CAIB Report, Vol. 1, Government Printing Office Washington, DC, 224 (August 2003).

[15] Risk and Lethality Commonality Team, Range Safety Group, Range Commanders Council, Common Risk Criteria for National Test Ranges: Inert Debris, RCC 321–02, White Sands Missile Range, New Mexico, 2–1 (2002).

[16] Air Force Space Command Manual 91–710 (AFSPCMAN 91–710), Range Safety User Requirements (July 2004).

[17] Federal Register, Part III, Department of Transportation, Federal Aviation Administration, 14 CFR Parts 413, 415, and 417, Licensing and Safety Requirements for Launch; Final Rule, Vol. 71, No 165, 50537 (25 August 2006).

[18] This range of estimates was due to uncertainty in the fraction of people outside of a type of shelter and the amount of debris that survived to impact.
[19] M.Y.Y. Lin, E.W.F. Larson, and J.D. Collins, "Determination of Debris Risk to the Public Due to the Columbia Breakup During Re-entry", Report No. 03-517-01, ACTA, Inc., Torrance, CA, prepared for Columbia Accident Investigation Board (September 2003). Also Appendix D.16 to the Columbia Accident Investigation Board Final Report, Vol. II, 475–506, 501.
[20] *Ibid.*
[21] Bryan O'Connor, Associate Administrator for Safety and Mission Assurance, quoted in article by T. Halvorson and J. Kelly, "Re-entry Paths Must Be Reassessed", Florida Today (28 June 2003).
[22] NASA Procedural Requirements (NPR 8715.5), Range Safety Programme (15 July 2005).
[23] *Ibid.*, 13.
[24] S.L. Carbon and E.W.F. Larson, "Modelling of the Risk to Aircraft from Space Vehicle Debris", Proceedings of the AIAA Atmospheric Flight Mechanics Conference Exhibit, San Francisco, CA, USA (August 2005).
[25] R. Van Suetendael, et al., "Accommodating Commercial Space Operations in the National Airspace System," Journal of Air Traffic Control, Vol. 47, No. 3 (July-September 2005). Also R. Van Suetendael, "Safety Considerations for Aircraft and Space Launch/Re-entry Vehicles Operating in Shared Airspace," The Journal of Air Traffic Control, Vol. 44, No. 1 (January–March 2002).
[26] R. Van Suetendael, et al., "Accommodating Commercial Space Operations in the National Airspace System," Journal of Air Traffic Control, Vol. 47, No. 3 (July–September 2005).
[27] P. Wild, R. Van Suetendael, J. Hallock, and E. Larson, "Public Safety Standards for the Launch and Entry of Spacecraft," 1st IAASS Safety Conference, Nice, France (October 2005).
[28] R. Van Suetendael, et al., "Accommodating Commercial Space Operations in the National Airspace System," Journal of Air Traffic Control, Vol. 47, No. 3 (July–September 2005).
[29] R. Van Suetendael, "Safety Considerations for Aircraft and Space Launch/Re-entry Vehicles Operating in Shared Airspace," The Journal of Air Traffic Control, Vol. 44, No. 1 (January–March 2002).
[30] M.Y.Y. Lin, E.W.F. Larson, and J.D. Collins, "Determination of Debris Risk to the Public Due to the Columbia Breakup During Re-entry" Report No. 03-517-01, ACTA, Inc., Torrance, CA, prepared for Columbia Accident Investigation Board (September 2003). Also Appendix D.16 to the Columbia Accident Investigation Board Final Report, Vol. II, 475–506.

3.6 Saving lives in space missions

3.6.1 Extending international search and rescue

The 1912 Titanic disaster, with a distress message telegraphed in Morse code, was a defining moment in starting the organization of search-and-rescue means on a global scale. The shock of that shipwreck and the new public awareness that it bred led to the establishment of means for constant distress surveillance on land and aboard ships. The Titanic disaster led in 1914 to the first International Convention for the Safety of the Life at Sea (SOLAS),[31] which required ships to go to the assistance of other vessels in distress. The system developed and matured gradually in the following decades and in the early 50s it was extended to aviation, but it was only in 1985 that a well organized international search and rescue (SAR) system came into force under the International Convention on Maritime Search and Rescue of 1979.[32] The current international SAR system is based on close coordination between IMO and ICAO, and relies on uniform worldwide coverage and use of global space-based monitoring and tracking resources available on board GEO and LEO spacecraft (COSPAS-SARSAT Programme).

As with any comparable system, the safety of crew and passengers on board future suborbital and orbital commercial space vehicles will not depend only on design adequacy, robustness of construction and the capability to tolerate failures and environmental risks, but also upon special provisions which may allow escape, search, and timely rescue in case of emergencies. During a suborbital commercial human spaceflight, an emergency may lead to search and rescue operations at sea or on land not dissimilar from those of an aviation accident. The case of an on-orbit emergency is different and for that, special cooperation provisions and interoperable means need to be developed. Here, the closest parallel is that of submarine emergencies. Many nations now regularly practice multilateral rescue exercises and coordinate their rescue means and capabilities through the International Submarine Escape and Rescue Liaison Office (ISMERLO).

3.6.2 Ascent emergencies

During the ascent phase, a so-called abort scenario needs to be considered in order to safeguard the life of the crew and passengers on board a commercial space vehicle. Such scenarios apply to any vehicle configuration (winged or capsule) and would require in several cases planning and cooperation with foreign countries.

Taking the experience of the Shuttle Programme as an example, it will be seen that depending on the time a malfunction occurs, there are Shuttle international launch abort sites at Halifax, Stephenville, St. Johns, Gander, and Goose Bay (all in Canada). There are also Shuttle transoceanic abort landing sites (TAL) at Ben Guerir Air Base, Morocco; Yundum International Airport, Banjul, The Gambia; Moron Air Base, Spain; Zaragoza Air Base, Spain; and Istres, France. Finally, there are eighteen designated Shuttle emergency landing sites spread amongst Germany, Sweden, Turkey, Australia, and Polynesia, several of which are active international airports. For the purpose of providing the Shuttle programme with the necessary assistance, access, and dedicated capabilities at those foreign landing sites worldwide, the U.S. government had to negotiate a large number of specific bilateral agreements, thus implicitly recognizing that the 1968 Rescue and Return Agreement, even when signed by the other governments, is not detailed and sufficiently precise enough for this purpose.

In the future, when commercial human suborbital and orbital spaceflights become common, private entities will be able to gain the same level of assistance on land or at sea, and access to foreign facilities only if the necessary international civil space agreements and regulations are put in place by an international space regulatory body.

3.6.3 Orbital safety and rescue

The idea that space-faring countries should take measures to make mutual aid possible in case of on-orbit emergencies is about twenty years old. In July 1990, an International Spacecraft Rendezvous and Docking Conference was held at the NASA Johnson Space Centre. The Conference was initiated by NASA at the direction of the U.S. Congress. During the introductory plenary session, J. Loftus of NASA gave a background for the Conference, referencing the numerous discussions that had previously taken place in the IAF's Safety and Rescue Symposia, the 1968 Rescue and Return Agreement, and the uncountable number of appeals made by the Association of Space Explorers. The purpose of the meeting was to explore the need for international consensus to establish a set of common space systems design and operational standards which would allow docking and on-orbit interoperability in case of emergency.

The attributes for such international standards were summarized at the Conference as follows:

- Each party could implement them with their own systems and resources;
- Cooperation in such standards does not require subordination (i.e., one party does not have to buy parts of the system from another);
- Success of one project or project element is not required to insure success of the other;
- No one standard requires subordination to another standard;
- The functional requirements of the standard can be implemented with a number of alternative technologies. Definition of the standards does not require the transfer of technology.

Some participants at the Conference commented that such rendezvous and docking standards "would eventually have a similar role to that of the international civil aviation certification standards which permit civil aircraft designed and manufactured in one nation to operate within the airspace, and to land at the airports, of other nations".[33]

It should be noted that the Conference was intended to build upon the wealth of experience gained in previous cooperation between the two key players of the time (i.e., Russia and the U.S.), which had developed the so-called androgynous docking system APAS as part of the Apollo-Soyuz Project, later used on the Shuttle and ISS. With the termination a few months after the Conference of the "mini-shuttle" programmes in Europe and Japan, and with the Russians joining the International Space Station Programme, interest in the development of the above-mentioned common international rendezvous and docking standards quickly waned.

Nowadays, the enlargement of the manned space-faring club to include players like China, India and, perhaps one day, Europe and Japan, raises to the forefront once again the issue of international interoperability space safety standards. In fact by the NASA Authorization Act 2008, the U.S. Congress directed the NASA Administrator to enter into discussion with space-faring nations to agree upon a common docking system standard for crew rescue from stranded spacecraft.[34] But there is another reason why such common international standards for spacecraft on-orbit interoperability, rendezvous, and docking may emerge sooner than later. It is the so-called Commercial Orbital Transportation Services initiative that is being pursued by NASA for purposes of uploading cargo to the International Space Station.[35] As part of this initiative companies will be able to provide commercial cargo transportation services to the ISS initially with automatic vehicles, and possibly with crew capabilities in the future.

3.6.4 Ionizing radiation risk for human spaceflight

The Earth's magnetic field traps electrically charged radiation particles in two belts high above the Earth. The highest extends out to about 40,000 km, and the lowest belt begins at about 600 km above the surface. The intensity of radiation in these belts can be more than a million times higher than on the Earth. For several decades to come, commercial manned orbital spaceflight will most probably be limited to low Earth orbit flights where the radiation level is small or negligible. Based on the experience of several decades of manned spaceflight in low Earth orbit, a safe level of radiation exposure has been defined as that which would increase the lifetime risk of cancer by 3%, and this translates into a total dose of 100–400 rem depending on age and gender. For comparison, a maximum of 10 rem is the annual dose allowed for workers in occupations involving radiation. Since health risk increases with the total dose, it is important to monitor the dose and to establish norms for the retirement of (commercial) astronauts who reach that level.

[31] The text of the SOLAS Convention is available online: http://www.derechomaritimo.info/c-solas1974.htm (last accessed: 03 January 2011).

[32] Available online: http://www.admiraltylawguide.com/conven/searchrescue1979.html (last accessed: 03 January 2011).

[33] Internal ESA/ESTEC-QSS Memo from K. Wright Head of Safety Section to ESTEC Director M. Lefevre, Noordwijk, the Netherlands (September 1990).

[34] Public Law 110–422 (section 407) Oct. 15, 2008. Available online: http://legislative.nasa.gov/PL%20110-422.pdf (last accessed: 03 January 2011).

[35] http://www.nasa.gov/offices/c3po/home/cots_project.html (last accessed: 03 January 2011).

CHAPTER 4

NEED FOR INTERNATIONAL SPACE SAFETY REGULATIONS

4.1 Need for international regulation of STM, space tourism & space debris

4.1.1 Commonality or un-commonality of ground standards

On 15 October 2002, a Russian Soyuz launcher exploded some twenty seconds after lift-off from the Russian Plesetsk cosmodrome. The launcher's payload was an unmanned Foton M-1 research satellite containing 44 experiments belonging to the European Space Agency. One of the four Soyuz boosters malfunctioned after launch and lost power. It fell away from the vehicle as it is designed to do when thrust no longer holds it in place and upon impact with the ground its tanks ruptured causing a large fire that resulted in significant damage to the pad. The launcher then automatically shut down the three other boosters about twenty seconds into the launch and the entire rocket fell back, struck the ground, and exploded in a huge fireball at a location about 1 km away from the launch pad. Apparently, the supply of hydrogen peroxide to the propellant turbo pumps was blocked by a metallic object. The explosion killed a Russian soldier watching the launch from the first floor of the integration building. Fortunately, all forty engineers and scientists from various European countries involved in the preparation of the spacecraft who were also watching the launch from a location closer to the explosion but on lower ground were unharmed in the accident.

The Plesetsk accident exemplifies the fact that launch sites open to international payload launches potentially expose foreign personnel to the safety risks of vehicle processing and launch. In the same manner, it should also be noted that the preparation of a foreign payload exposes local ground personnel to the risks associated therewith. The obvious consequence is that, on the basis of mutual interest, the level of risk (from local launchers and foreign payloads) deemed acceptable at international launch sites and the relevant safety control measures should be identical and comparable at any launch site in the world, irrespective of where it is located.

For the payloads, such uniform ground safety requirements would have additional advantages in terms of economy, flexibility, and fair competition.

In fact, the adoption and implementation of common ground safety requirements would greatly reduce the design impact of switching launchers, should the need arise. Furthermore, common ground safety requirements and their uniform implementation worldwide would exclude variations in programme safety costs from the commercial competition equation.

In perspective, the introduction of uniform ground safety technical requirements and certification practices could possibly pave the way for a regime of mutual (i.e., bilateral or multilateral) recognition of the safety certificates that may be granted by a national safety authority as is presently the case in some other fields of endeavour (e.g. civil aviation). Such a regime would further enhance the overall safety system by not limiting safety certification exclusively to design aspects (as is the case today), but extending it to cover elements such as independent quality assurance surveillance of the actual processes of manufacturing and testing by the local certificating authorities in the manufacturer's country. In other words, such a regime would make it possible for payloads to be safety-certified "at the source", thus eliminating (or at least limiting) the need for disclosure of detailed proprietary information on the design and operational features of the payload to the safety authorities of foreign countries from where such payloads are to be launched.

4.1.2 Comprehensive regulatory approach to space traffic management

More and more space actors are coming to the stark realization that some of humanity's current activities in outer space are unsustainable in the long term and could result in degradation or complete loss of utility of certain orbits. To this end, some progress has been made in the field of space debris mitigation – a fundamental part of the broader process of ensuring the long term sustainability of space. However, space debris mitigation is only one piece of the puzzle as it focuses almost exclusively on limiting the creation of new debris. It does not deal with the risk posed by the tens of thousands of pieces of existing debris, nor does it address the interaction between such debris and operational spacecraft.

Space traffic management (STM)[1] is another important means of complementing debris mitigation efforts and enhancing the long term sustainability of space. STM has been a subject of serious reflection since the 1980s. To date, the most in-depth work on STM was undertaken by the International Academy of Astronautics (IAA)[2] which published a "Cosmic Study" on that topic in 2006. This 100-paged work, prepared by an international study team of the IAA, aimed at providing a multidisciplinary status report on technical, regulatory, and policy aspects of STM.[3] Additional work on the subject carried out during the 2007 Summer Session of the International Space University led to the production of a report that built on the Cosmic Study and made the first attempt at drafting a rough set of STM rules.[4]

Perspectives on exactly what constitutes space traffic management differ widely between various institutions, experts, and studies. However, for our present purposes, space traffic management encompasses the entire gamut of technical and regulatory guidelines and requirements necessary for the prevention of unintentional physical or radiofrequency interference in the conduct of space activities. STM is required in order to promote safe, efficient and sustainable access to and use of, outer space.

While comprehensive and coherent regulations have been developed for proper traffic management in the maritime and air transport domains, space traffic management is still very much at an *ad hoc* stage. Some elements of STM are already in existence, but they are by no means widespread or uniform among the broad majority of space actors nor is there a comprehensive legal regime in place to enforce the implementation of these measures.

For example, the United States Military conducts daily screenings of all active and manoeuvrable payloads in Earth orbit for close approaches (called conjunctions) with other objects.[5] In the event that a conjunction is found, the United States will warn the owner-operator of that particular satellite who then makes a decision on whether or not to manoeuvre it to avoid collision. Additionally, a number of commercial satellite operators have formed the Space Data Association (SDA) and have tasked it with the responsibility of combining and analyzing positional data about their respective satellite fleets and providing a conjunction assessment/collision avoidance service for participating operators as well as helping them to resolve radiofrequency interference issues.[6] Other satellite operators such as the European Space Agency and Russian military also perform their own conjunction assessments and collision avoidance for satellites under their control.

Unlike space debris mitigation, the perception of a real need for a comprehensive space traffic management regime is currently not shared by all of the actors who are actively conducting space activities. Just as with the other global commons, a comprehensive STM regime will only be established when these actors (public, private, military and civilian) come to an understanding of the benefits of STM to the extent that they will accept the restrictions on their freedom of action in space that an STM regime would necessarily impose. Thus, it is important to find the means to increase all stakeholders' current level of understanding concerning the need to regulate certain space activities and the potential long-term benefits of doing so at this critical point in time, before the problems grow past the point of no return.

4.1.2.1 Legal basis for potential STM regime

In developing any potential STM regime, there are certain elements which must be included if the proposed regime is to be of any practical value. The first is that it must be rooted within the principles of existing international space law. First and foremost is the principle, laid out in the 1967 Outer Space Treaty, that all countries have an equal right to carry out space activities in accordance with international law.

A second fundamental issue that any prospective STM regime must deal with is the long tradition of freedom of action in outer space and the vital role that space plays in the national security of many States. States are accustomed to having almost entirely free rein in conducting their space activities, and over the last several decades they have, in reliance upon this freedom of action, deployed a large number of expensive and sensitive space assets which are critical to their national security. Underlying this issue is the changing nature of the space domain. Fifty years ago, space was essentially the exclusive domain of two super powers (the United States and the Soviet Union) who were then engaged in the decades-long struggle of the Cold War. During that time, these States developed an operational familiarity with each other in space. Both also recognized that possession of space assets was a crucial element of nuclear deterrence for purposes of reducing the risk of nuclear war. As such, the space treaties and unwritten understandings that were developed during that era were properly suited to regulating the activities of the super powers in space.

The current security regime in space is very different. While the United States remains the most prominent actor possessing the most space assets, there are now about 50 States operating satellites (11 of which have indigenous capability to place objects in space) and a host of multinational private entities and non-State actors actively exploring and exploiting space. Recent events have demonstrated that certain actions carried out by any one of these actors can have negative impacts on all actors and the space environment as a whole. Complete freedom of action by all these actors is increasingly being seen as detrimental to the continued long term sustainability of access to and use of outer space by all.

This issue of freedom of action in the airspace and its national security implications was also a matter of concern during the creation of the global regime of air traffic management. The 1944 Chicago Convention which forms the basis for international air traffic management addressed this issue in its Article 3:

(a) This Convention shall be applicable only to civil aircraft, and shall not be applicable to State aircraft.
(b) Aircraft used in military, customs and police services shall be deemed to be State aircraft.

(c) No State aircraft of a contracting State shall fly over the territory of another State or land thereon without authorization by special agreement or otherwise, and in accordance with the terms thereof.
(d) The contracting States undertake, when issuing regulations for their State aircraft, that they will have due regard for the safety of navigation of civil aircraft.

This Article established the critical distinction between State aircraft and non-State aircraft and, most importantly, excluded State aircraft (including those operated by the military) from the scope of application of the provisions of the Convention. State aircraft are, however, required to operate with *due regard* for civil and commercial air traffic, and the State would be held liable for actions which compromise the safety of civil and commercial air traffic. The compromise brought about by this Article removed a significant road block that would have otherwise seriously jeopardized the successful negotiation and adoption of the Convention. As a result of this Article, State aircraft are strictly speaking not required to follow air traffic rules. In the vast majority of cases, however, they do because it is in their self interest to coordinate for their own safety. ICAO attends military committee(s) of NATO to discuss such policy.

It should, however, be noted that Article 3 does not achieve this result on its own. It acts in combination with Article 1 and other Articles to allow States to designate specific portions of their airspace for the use of civil and commercial aviation. In most cases, the remainder of a State's airspace is considered military airspace and special permission is required before any civil or commercial aviation activity can take place there. This makes it easy to keep State aircraft physically separated from non-State aircraft; i.e. they fly in physically separate portions of the airspace. In practice, the same may be difficult to achieve in outer space due to the operation of Article I of the Outer Space Treaty, which prohibits the exercise of sovereignty in any form whatsoever. Therefore, it will be necessary to conclude an international agreement under which States could undertake to keep their civil and commercial spacecraft physically separated from their respective State (military) spacecraft. In fact, to do so it will be imperative for States to have in place an international STM system established pursuant to a comprehensive approach towards the STM regime.

Including a similar exception for State spacecraft is perhaps even more important to the success of any STM regime because of the long standing importance of freedom of action in space to the national security of many States. And just as is the case with air traffic, the vast majority of State spacecraft will voluntarily follow the rules because it will be in the self-interest of States to do so.

The third major requirement for any proposed STM regime is that it must be technically feasible. This may seem like an observation too obvious to mention, but there have been cases in the past where STM regimes were proposed on the basis of air or maritime traffic management regimes without taking into account the unique technical requirements and constraints of outer space.

For example, the maritime traffic management regime stipulates that "Every vessel shall at all times proceed at a safe speed so that she can take proper and effective action to avoid collision".[7] While this rule is completely feasible in the maritime environment, it is entirely unworkable in space. The speed at which a satellite travels is not an independent variable – it is linked to the altitude at which the satellite orbits. Any change in the speed dictates a change in the altitude, and vice versa.

One way to create such technically feasible rules would be to model the process after that used for the development of the COPUOS Space Debris Mitigation Guidelines. These guidelines started as a series of technical measures which were drafted over a number of years by the Inter-Agency Space Debris Coordination Committee (IADC), an organization composed of technical experts from those States with active space debris research programmes. Another potential way forward would be to work from a common set of established best practices which could then be eventually codified into an STM regime.

The fourth major element that must be observed in any proposed STM regime is that it must be adopted primarily for the civil and commercial use of outer space and not negotiated through a forum for fostering space arms control or disarmament. This is mainly due to political reasons and for ease of adoption. Within the United Nations system, the civil and commercial use of outer space (including the space debris problem) is dealt with by COPUOS. The arms control and disarmament aspects of outer space are discussed within the Conference on Disarmament (CD), a multilateral negotiating body located in Geneva, Switzerland. Within the CD, space issues are part of the prevention of an arms race in outer space (PAROS) initiative, which is linked to the other arms control issues of nuclear disarmament, control of production of fissile material, and assuring non-nuclear weapon States against the threat or risk of nuclear weapons. Any proposed STM regime should maintain this clear and distinct separation between the peaceful uses of outer space and arms control issues that has been maintained within the UN system. Otherwise, it will require discussion and achievement of consensus of both COPUOS and the CD, a monumental task.

Finally, any prospective STM regime should focus on enhancing the safe and efficient use of space by all actors and the long term sustainability of Earth orbit without adding undue restrictions that stifle innovation and commercial development. Space should not be persevered simply for the sake of keeping it pristine –

the object of long term sustainability must be pursued for the express purpose of maintaining the ability to use space for its many benefits.

Additionally, any STM regime that is perceived as too onerous will not be practical and will have the unintended effect of limiting the potential for use of space for benefit on Earth. Therefore, it is recommended that any proposed STM regime must begin by tackling the most serious problems, such as on-orbit collisions, end-of-life disposal, and radiofrequency interference and, over time, gradually add additional regulations to cover other aspects of space operations.

In brief, it can be said that a comprehensive approach to satisfy the need for sustainable use of space by all States and non-State actors is highlighted in the concept of STM. Space activities are approaching a stage where comprehensive and coherent traffic management regulations comparable to those that pertain in the realm of air traffic management should be envisaged. While already there are elements of STM in existence (e.g., those applicable in the geosynchronous orbit), a comprehensive regime does not exist. Also, the actors who are actively conducting space activities do not currently share the perception of a real need for such a system. An STM approach will only be established when these actors (public or private, military or civilian) see the benefits of restricting their freedom of action through specific traffic management rules. It is important to increase understanding about the need to comprehensively regulate space activities at this critical stage when the problems have not become too pressing and difficult to resolve.

The general principles of space law provide a basis and rationale for establishing a STM regime. However, current international space law lacks provisions essential to the creation of a comprehensive traffic management regime (i.e., pre-launch notification). Of particular importance is legal recognition of the difference between space objects considered as valuable assets by their owners, and space debris that has no value.

The implementation of a comprehensive STM regime would require additional regulations (with regard to the provision of information about, and the execution of, space missions), which could be perceived by States as limiting their freedom of use of outer space guaranteed by the Outer Space Treaty. In order to achieve consensus, States must appreciate the existence of a certain level of urgency and must also expect specific as well as collective benefits similar to those accruing from existing regulations.

4.1.2.2 Comparable traffic management regime

In international common spaces such as outer space, the high seas, and international airspace, no territorial sovereignty and control applies. However, within sovereign national airspace, air traffic management (ATM) is within the exclusive preserve of the national authorities concerned. As much as possible, these authorities apply the uniform "standards and recommended practices" (SARPs) developed by ICAO. Aircraft commanders are accustomed to being transferred from one national ATM agency to another, either on the basis of horizontal border crossing or because their aircraft has reached a certain altitude, and they routinely receive instructions from the ground as to recommended speed, altitude, routing, and right of way. The ATM agencies' responsibility for aviation safety also requires coordination with military and with space agencies. These national responsibilities may be delegated to another State or to an international agency. All ICAO rules also apply to aircraft in airspace over the high seas, a *res communis* like outer space. A national ATM agency may consent with ICAO to take responsibility for a specific portion of the airspace over the high seas. The ICAO ATM system is highly sophisticated and effective and may serve as an example of rule-making for STM purposes.[8]

4.1.2.3 Other considerations for setting up an STM system

In order to set up an international operational STM system, several questions and perspectives need to be addressed. In this regard, an analysis of space activities in the current framework leads to numerous insights and conclusions that amply demonstrate the fact that further research is needed on the following:

- A study of the experience gained by those countries which have already established national licensing systems for commercial space operations should be undertaken;
- A study should be conducted on the potential negative effects certain space activities may have on the space environment (such as funerals or advertising displays), including whether certain space activities or specific uses of outer space should be regulated or, perhaps, even banned in order to protect long-term sustainability;
- Ways of linking/merging the ITU information/notification system with an improved UN registration system, the outcome of which would be a unified international notification/information system should be found;
- The relationship between international space law and the Hague Code of Conduct against Ballistic Missile Proliferation (HCOC) regarding the concept of notification of launches should be examined;
- Further enquiries into the interests and expectations of private actors and the economic benefits of STM to commercial activities should be carried out;

- The latest trends in technical international organizations like ITU, ICAO, or IMO regarding the adoption of technical regulations/standards should be studied so as to provide more flexibility than the traditional system of negotiation and ratification currently provides. These trends should be analyzed in connection with their relevance to an STM regime; and
- Finally, various policy initiatives for maintaining common safety standards, as well as avoiding "flags of convenience", through STM should be proposed.

4.1.3 International regulation of aerospace vehicles for space tourism

During its development flights, SpaceShipOne received a flight permit from the U.S. Federal Aviation Administration (FAA) as a glider. For flight operations, SpaceShipOne was granted a launch license by the FAA Office of Commercial Space Transportation (AST) as a "Reusable Launch Vehicle" (RLV), classified as a rocket. SpaceShipOne raises various international and domestic regulatory questions:

a) How should hybrid vehicles like SpaceShipOne be categorized? As a rocket? As an aircraft?
b) Do we need to create new classifications?
c) Do we need "integrated" aero-space worthiness regulations and certification teams?

Annex 7 to the Chicago Convention contains an internationally agreed upon definition of aircraft, namely: "Any machine that can derive support in the atmosphere from the reactions of the air". This definition has its origins in an Annex promulgated in the 1920s by ICAO's predecessor, ICAN, established by the Paris Convention of 1919. Back in the 1920s, it was only the students of the writings of Jules Verne who contemplated that one day vehicles would lift men into space. A spacecraft, on the other hand, is defined under U.S. national law as "A man-made vehicle which is intended to go beyond the major portion of the Earth's atmosphere".[9] A rocket is defined as "(A) a vehicle built to operate in, or place a payload in, outer space; and (B) a suborbital rocket".[10] The American Space Shuttle, though it acts like an aircraft (a glider) on its descent, has been classified and treated as a spacecraft under U.S. national law.

SpaceShipOne has wings and aerodynamic surfaces which are essential during its re-entry into the atmosphere when it operates as a glider. During the ballistic portion of its flight, the air only contributes to aerodynamic control of the vehicle from launch altitude until it approaches the 100 km altitude where the air density is no longer sufficient for aerodynamic flight control. Consequently, for certain parts of the flight and for most of the time, SpaceShipOne operates as an aircraft.

How then are the other portions of the SpaceShipOne's journey to be classified? First of all, for a large part of its ascending phase, SpaceShipOne was the external winged payload of a carrier airplane. Because it influenced the overall aerodynamic configuration, flight characteristics, and structural loads, the vehicle was also a key part of the carrier aircraft's airworthiness certification. The question is whether SpaceShipOne, and its successor SpaceShipTwo, can be considered space vehicles because of the few minutes during which they fly in space.

Classifying a system performing suborbital flights as an aircraft, a space vehicle, or something in between is, in certain cases, difficult because of the lack of standard definitions. However, such classifications have potentially significant regulatory and legal repercussions (aside from safety concerns) which may have an important impact on the emerging space tourism industry. Certain rules may become applicable to the vehicle as a result of the classification adopted. For example, if a suborbital vehicle is to be classified as an aircraft when engaged in international point-to-point transport, then it would be subject to the provisions of the Chicago Convention in terms of registration, airworthiness certification, pilot licensing, and operational requirements.

On the other hand, inappropriate classification could lead to the certification of vehicles which do not completely meet the minimum safety standards expected of state-of-the-art vehicles. This may occur as a result of the application of inappropriate design and testing standards, or because the certification team lacks the skills and competencies necessarily required to properly certify such vehicles. Finally, there may be resistance to call things by their proper names because of marketing strategies. Selling a ride on a supersonic (or even hypersonic) aircraft flying to an altitude of 100 km does not have the same appeal to potential wealthy customers as marketing the same ride as an opportunity to undergo and experience what astronauts do, although this will only occur at the edge of space.

In conclusion, when determining the applicable regulations, consideration should be given to the "prevailing" functionalities of the vehicle, and also to the safety of the public on the ground and that of the passengers and crew on board. The applicable regulations should be applied coherently and consistently, including difficult aspects such as the implementation of flight termination systems. In other words, the level of system safety should be driven by the accumulated knowledge and best practices in the most closely related field(s). The classification of vehicles, and consequently the selection of an appropriate safety certification regime, should not be influenced by the desire to promote the initial development of a new industry irrespective of the legitimacy of those desires. At the same time, the regime should be devoid of any penalizing aspects not suited to the technological environment. It appears that ICAO's involvement in both sectors will become imperative and unavoidable as space activities (particularly aerospace flights) expand, and interaction between aviation and space activities continues to increase.

4.1.4 International regulation of orbital debris

In 2002, after a multi-year effort, the IADC adopted a consensus set of guidelines designed to mitigate the growth of orbital debris. In 2006, the Scientific and Technical Sub-committee of COPUOS adopted space debris mitigation guidelines, essentially based on the IADC guidelines, which were subsequently endorsed by the UNGA in 2007.

Neither the IADC nor the COPUOS guidelines are legally binding under international law and it is also recognized that exceptions to the implementation of individual guidelines or elements may be justified. The relative value of the COPUOS guidelines was brought to the limelight by the Chinese ASAT test of January 2007, which created a new massive and hazardous amount of orbital debris. Although China had actively participated at all stages of the development and adoption of the guidelines, its ASAT test was not in conformity with many of the debris mitigation guidelines contained therein. This has renewed doubts as to the legal efficacy and effect of the guidelines as adopted by the COPUOS.

It is therefore recommended at the international level that COPUOS should start discussing the question as to whether or not pieces of space debris fall within the meaning of "space objects" as used in the existing space law treaties. If a space object is deemed "worthless", should it be considered abandoned by the owner and/or should salvage rules apply? If it is decided that space debris are not space objects then additional protocols should be elaborated stating what provisions of the treaties apply to valuable spacecraft and which provisions apply to space debris. These protocols should determine under what conditions space debris may be removed or re-orbited in order to prevent collisions or close encounters with valuable spacecraft.

Both the IADC and the COPUOS guidelines are an important baseline for further national initiatives. Acting thereunder, a number of governments have issued national guidelines and have also established formal processes for reviewing new space missions and their potential for on-orbit debris generation. In October 2005, the Federal Communications Commission (FCC), the primary U.S. agency responsible for authorizing non-governmental satellite activities, announced that any applicant with a pending request for authority would be required to file an orbital debris mitigation plan, and all new license requests would require filing of the same plan.

A further important initiative is underway at the International Standards Organization (ISO). The Orbital Debris Co-ordination Working Group (ODCWG), set up in 2003 after initial ISO discussions, has been responsible for developing a plan for the preparation of standards on debris mitigation, and for managing the preparation of those standards by the technical sub-committee on "Space Systems and Operations" (known as ISO TC20/SC14). The ISO resolutions which initiated the relevant standardization activities recognized three key points:

- International dimension of the space debris mitigation issue;
- Comprehensive and cohesive system of implementing standards is essential to achieve orbital debris mitigation objectives and
- While it is the ISO's responsibility to develop internationally recognized standards on a consensus basis, it is up to the governments to ensure their implementation through national or international regulations.

There are six standards projects in development, and a further seven project proposals are being prepared. International standards, technical specifications, or technical reports are expected to be published in 2011 or 2012.

Finally, because all the above-mentioned practices are only voluntary guidelines and not binding regulations, there is no legally enforceable responsibility to implement them and/or accountability for their implementation. More importantly, it is well recognized that space debris is a serious navigational hazard for space missions, threatens people on the ground, and could pose risks for aviation as well. Therefore, it is important that uniform technical standards, binding regulations, and procedures to regulate this hazard be developed by an international regulatory body, like ICAO.

---

[1] There have been many discussions over terminology, including whether the term "Control" should be used instead of "Management". For the sake of brevity, this document will use Management.
[2] http://iaaweb.org/iaa/Studies/spacetraffic.pdf (last accessed: 03 January 2011).
[3] http://iaaweb.org/iaa/Studies/spacetraffic.pdf (last accessed: 03 January 2011).
[4] http://www.isunet.edu/index.php?option=com_content&task=view&id=374&Itemid=251 (last accessed: 03 January 2011).
[5] Staff Sgt. Benjamin Rojek. "JFCC-Space achieves flight safety milestone", 1/14/2010. Online: http://www.vandenberg.af.mil/news/story.asp?id=123185586 (last accessed: 03 January 2011).

[6] http://www.space-data.org/sda/ (last accessed: 03 January 2011).
[7] Convention on the International Regulations for Preventing Collisions at Sea, 1972, Rule 6. Available online: http://www.admiraltylawguide.com/conven/collisions1972.html (last accessed: 03 January 2011).
[8] See, P. Dempsey and M. Mineiro, "ICAO's Role in Regulating Safety and Navigation in Suborbital Aerospace Transportation", presented at the 2008 IAASS conference: "...while the term 'airspace' and 'outer space' are not clearly defined, any activity whether 'space' related or 'air' related that occurs or affects international civil aviation, in particular when the activity occurs in the medium of 'airspace' that is traditionally utilized by civil aviation, requires co-ordination by ICAO".
[9] U.S. Code of Federal Regulations – Title 47: Telecommunication, Chapter I, Part 25 (Satellite Communications), Subpart C – Technical Standards, Sec. 25.201 (Definitions); 47CFR25.201.
[10] U.S. Code – Title 49: Transportation, Subtitle IX (Commercial Space Transportation), Chapter 701 (Commercial Space Launch Activities), Sec. 70102 (Definitions), (7).

CHAPTER 5

PROPOSAL FOR A NEW REGULATORY REGIME

5.1 ICAO for near-space safety?

5.1.1 Background

In this Chapter, several other factors and developments are examined to further assess the rationale for ICAO's involvement in space. Also included are the modalities for implementing the ICAO for Space proposal.

The idea of an international organization for space modelled along the lines of ICAO is not a new one. It was put forward several years ago by FAA-AST officials under the name of *International Space Flight Organization* (ISFO). At the time, it was envisaged that the organization would focus on futuristic international traffic of commercial hypersonic space planes, and on the launch of space vehicles from one country which, on its return to Earth, would land in a foreign country. Clearly, potential interest in such operations, and therefore in the ISFO concept, was limited to a handful of countries. This Study proposes the establishment of such an international flight organization based on the international nature of the safety and environmental risk of space missions, no matter where they originate or end, and the fact that the risks are destined to increase dramatically as space activities expand and more countries gain access to space technologies. An international regulatory organization is therefore needed to

ensure that proper steps are taken at this time to assure sustainable access to and use of space. This is underscored by the fact that all nations (whether currently active in space or not) have an actual or potential stake in the commercial and military utilization of the near-Earth space and, by necessary implication, in its preservation as an asset for all mankind.

This part of the Study advances various reasons and underlying rationales aimed at demonstrating that the essential functions expected to be performed by an international space flight organization would be better and more efficiently carried out under Standards and Recommended Practices promulgated by ICAO. In short, instead of establishing a new international space flight organization, it is argued that the same ends could be achieved, as a starting point, simply by extending the mandate of ICAO to the region of space up to and including the geosynchronous orbit. Realistically, this is the region of commercial interest for the next half a century. Eventually, if and when in future appropriate technology makes it possible for outlying regions beyond the geosynchronous orbits to become routinely usable for commercial space operations, ICAO's mandate might again be extended thereby entitling it to be responsible for the regulation of all forms of operations and transport in space. In brief, an evolutionary approach to the expansion of ICAO's mandate is advocated.

The Study first revisits the historic "cultural" links between the two main (sometimes competing) interests in the near space region (i.e., aerospace military forces and commercial space operators). Following that, the Study demonstrates how existing aviation regulations would naturally overlap with future space regulations, and discusses the trend toward sharing and integration of ground and on-orbit infrastructure. Finally, this part of the Study makes the point that the organizational efficiency and the deeply rooted safety culture practiced at ICAO would bring immense benefits to the space community if the ICAO for Space proposal is carried through.

5.1.2 Management of space-bound traffic through international airspace

The oceans and seas are very important for the safety of space launch and re-entry operations. Major spaceports and launch sites are thus generally located close to the ocean coastline for the obvious safety reason of avoiding or fast clearing inhabited areas. In some cases, launches even take place directly from modified, self-propelled ocean oil-drilling platforms in order to provide for the most direct route to orbit and maximum lift capacity. As much as possible, spacecraft re-entry trajectories are selected on the basis of similar criteria, and all controlled destructive re-entries are directed to the oceans.

The Outer Space Treaty expressly prohibits the exertion of national sovereignty by any means whatsoever in outer space. On the other hand, as a result of the operation of the principle of complete and exclusive State sovereignty over national airspace under Article 1 of the Chicago Convention, the nationality of airspace is determined in accordance with the extent of the subjacent landmass and territorial waters under each State's sovereignty. More precisely, the territorial sovereignty of States reaches out to the "territorial sea" which extends a mere 12 nautical miles (nm) from the coast. There are then the so-called "contiguous zones", which are at 24 nm, in which each nation has the right to exercise control for purposes of preventing infringements of its customs, fiscal, immigration, and sanitary laws. Finally, the "exclusive economic zones" (EEZ) are defined as extending up to 200 nm from the baseline, and, in this zone, a nation controls the natural resources of the water and seabed. Beyond that line, the "high seas" begin. States have complete and exclusive sovereignty in their airspace which is defined as the atmospheric zone directly above their landmass and territorial sea. All the remaining worldwide airspace (i.e., the airspace above the contiguous sea zones, EEZs, high seas, and the Antarctic) is international. The Chicago Convention of 1944 placed such international airspace under the regulation of ICAO. Thus ICAO is responsible to regulate safety and navigation over the high seas, covering some 72% of the totality of the Earth's surface. As such, the large majority of space-bound traffic takes place through the international airspace (i.e., airspace which is not under the sovereignty of any State), which falls under the purview of ICAO. Hence, precedent exists for ICAO regulation in areas devoid of State sovereignty.

The responsibility to provide air traffic control services in international airspace is allocated through ICAO to various countries, based generally upon factors such as geographic proximity and the availability of the required resources. Currently, the oceanic air traffic control system is procedurally based, relying heavily on flight plan data filed by aircraft commanders. There is no radar coverage over the ocean. Pilots must report their positions verbally or have them automatically sent through a relay station. The infrequency of position reports, coupled with limitations in navigational accuracy and communications, have resulted in the requirement of large spatial separation standards between aircraft. As aviation and space traffic continue to grow, ICAO has an increasing primary responsibility and duty of promoting innovative strategies to ensure the safety of the "integrated" air and space traffic in the international airspace, which is where the two types of traffic mostly interact.

5.1.3 Integration of aviation and space infrastructure

Today's air traffic management system for civil aviation is not very different from that of the 1960s. It is still fundamentally based upon radar tracking, reliance on analogue voice radios, and the guidance of air traffic controllers. In the future, civil aviation will make use of space-based systems for traffic management, approach, and landing. Such systems are currently under development and make use of Global Navigation Satellite System (GNSS) such as the American GPS, plus various precision augmentation systems and position broadcasting capabilities.

The Wide Area Augmentation System (WAAS), for instance, is needed to correct current GNSS signal errors caused by ionospheric disturbances, timing, and satellite orbit errors, and it provides vital integrity information regarding the health of each GPS satellite. The WAAS consists of ground reference stations that monitor GPS satellite data, master stations which collect data from the reference stations and create a GPS correction message, and the geostationary satellites which broadcast the corrected message to WAAS-enabled GPS receivers.

The Local Area Augmentation System (LAAS) focuses its GPS augmentation service on the airport area (approximately 20–30 miles in radius) for precision approach and departure procedures. If implemented, the LAAS will yield the extremely high accuracy, availability, and integrity necessarily required for Category I, II, and III precision approaches. Improved navigation accuracy in the cockpit will allow further applications such as the Automatic Dependent Surveillance (ADS) system. The ADS will truly revolutionize air traffic control by allowing aircraft to automatically broadcast their positions to various receivers (ADS-B mode) on other aircraft and on the ground.

Also, in the case of the safety system for monitoring rocket launches, there is a forthcoming transition from ground-based radar to GPS applications. One of the most important safety responsibilities of a launch range safety officer is to monitor the track of launch vehicles during flight and, in case of malfunction and risk to the public, to terminate the flight. The method used for flight termination depends on the vehicle being used, the stage of flight, and other circumstances of the failure. Propulsion is usually terminated and, in addition, the vehicle may be destroyed by on-board explosive charges in order to disperse propellants before surface impact, or it may be kept intact to minimize the dispersion of solid debris. Flight termination can also be automatically initiated by a break-wire or lanyard pull on the vehicle if there is a premature stage separation. It can be expected that for hybrid manned vehicles (aero-spacecraft), the preference will be to avoid the incorporation of flight termination systems that rely on explosive charges, but to rely instead on flight control redundancies as is the case presently for traditional aircraft. In such cases, the vehicles will make use of the same GNSS systems used in civil aviation for navigation and traffic control purposes.

In the near future, a number of critical aviation systems (including traffic control, high resolution weather forecasts, and digital aviation communications) will be based in space. This means that aviation safety will depend mainly upon the integrity and reliability of space-based systems and services. Assigning international coordination and control of the "near-space" region to ICAO instead of a separate international space organization would bring about obvious advantages in terms of synergies and efficiency.

Similarly, the already observed trend towards operating aero-spacecraft from dual-use ground infrastructure (airport/spaceports) requires a well integrated international regulatory framework both for flightworthiness and ground operations certification which a single organization (i.e., ICAO) would be better placed to perform in a more efficient manner than separate international space and/or aviation organizations.

5.1.4 Integrated terrestrial and space weather forecasts

As cross-polar aircraft traffic (>latitude 78°N) increases, the aviation industry is becoming concerned about a number of safety related issues such as disruption in High Frequency (HF) communications, navigation system errors (older Loral-C or current and future GNSS), risk of failure of avionics, and radiation dose hazards for crew. The same concerns exist for high altitude flights at latitudes above 50°N (e.g., space tourism).

With reference to navigation systems errors for instance, it is known that space weather influences the electrical distribution in the Earth's ionosphere, which in turn can significantly throw off the accuracy of GNSS units used for safety-critical applications. In principle, the GNSS uses known positions of satellites and their distances from a receiver to determine the location of the receiver. When charged particles emitted from the Sun arrive in the Earth's atmosphere, they can cause considerable variations in the electron density (number of electrons in a given volume) of the ionosphere, both in time and space. The ionosphere is the layer of the atmosphere extending upward from a height of about 80 km. Its tenuous gas is electrically charged enough to affect radio signals. The GNSS radio signals must pass through the ionosphere and in so doing they are subjected to variations in the electron density structure of the ionosphere. Changes in the electron density due to space weather activity can change the speed at which the radio waves travel, introducing a "propagation delay" into the GNSS signal. The propagation delay can vary from minute to minute, and such intervals of rapid change can last for several hours, especially in the polar and auroral regions. Changing propagation delays causes errors in the determination of the range, otherwise known as "range errors".

Solar radiation storms, also known as Solar Proton Events (SPE), can increase the risk of errors and failures in other safety-critical aircraft electronic equipment such as flight engine management computers. Extreme solar flares can cause complete HF radio blackouts lasting for hours. This is a particular safety concern for flights in those areas where the sole or main means of communications is HF radio (e.g., flights over Russian polar routes, Northern Canada and transoceanic routes).

The aviation and space weather specialist communities are currently soliciting ICAO to take the lead in coordinating international rules on space weather forecast dissemination and uniform risk mitigation responses.

5.1.5 ICAO for an improved international space safety culture

When the era of manned spaceflight started during the Cold War, the object of the "mission" was one of affirming national prestige. This mission evolved hand-in-hand with the relationship between the then two superpowers – U.S. and U.S.S.R. – from being a technology supremacy propaganda tool (up to the climax of the Moon landing) to become a tangible sign of political goodwill and mutual acceptance of the *status quo.* This culminated with the Soyuz-Apollo docking in the mid-1970s. Finally, after the collapse of the Soviet Union, spaceflight cooperation (i.e., the International Space Station) was used as a means to prevent a feared "migration" of technical skills towards rogue States.

In general, a notorious truth in the military codes and practices of almost all countries is that in time of war the accomplishment of the "mission" takes precedence over considerations of personal and even collective "safety". A good soldier is expected, if necessary, to sacrifice his life for the sake of mission accomplishment. As a consequence of the strong original imprint of the military and political attitudes of the Cold War on space programmes, mission accomplishment (and not safety) has been very much the driving force behind the development of space systems. Safety has been relegated to a position subordinate to the main military objective in driving the design and execution of space systems. Unfortunately, even during the Post-Cold War era the situation has not changed much, irrespective of the entry of private actors into space activities. Governments are now assuming the role of regulators. As a result, a country in which there are active private space actors would have some sort of safety regulation. But this sort of regulation, if in place, would normally apply to the private sector and not to government missions, which continue to use the "mission accomplishment" approach and are not meeting safety requirements on a priority basis.

On the contrary, there is no commercial industry in which the safety risk is treated as secondary to commercial goals, costs, or profits. The air transport, nuclear, and pharmaceutical industries are examples of industries in which there are deeply-rooted safety cultures. The public may accept certain risks as unavoidable (car accidents for example), but will not tolerate those failures which are within the reach of current knowledge and technologies to prevent, and which are caused by economic pressure or by a lack of sufficient management or regulatory attention. As a consequence, public acceptance of failures and risks may eventually dictate the fate of a business. There is no evidence that the general public would have more tolerance for accidents in space projects brought about, for example, by inadequate testing of a new commercial aero-spacecraft while hundreds of millions of dollars are spent on lavishly appointed tourist facilities. While a certain level of safety risk is unavoidable in complex manned civil exploration missions conducted by professional astronauts, the safety risk acceptance criteria for commercial space programmes cannot be based on economic trade-offs, but on best practices and state-of-the-art technical knowledge. In this respect, the extension of the mandate of ICAO to commercial space activities would provide an unequalled wealth of organizational experience in establishing an internationally encompassing and deeply rooted safety culture such as that which has been instrumental in making civil aviation the great success of which we are all aware. The wise words of Jerome Lederer, the father of aviation safety, aptly demonstrate the idea being put forward here: *if you believe that safety is expensive try an accident!*

5.2 Proposal for a new regulatory regime

5.2.1 Policy principle

As we have seen, there are three categories of people who are exposed to space safety risks:

1) The general public on the ground, in the air, or at sea, because of launch and re-entry of space vehicles and parts thereof, or because of malfunctioning space-based, safety-critical systems;
2) Professional astronauts on exploratory and scientific missions and
3) Private crew and passengers on commercial spaceflights.

A combination of factors raises the central question of how to ensure the safety of the public and space crew as well as the integrity of other valuable unmanned assets in-orbit. These factors include the ever increasing number of actors involved in civil and commercial launch and re-entry activities, the envisaged expansion of human access to space for tourism/travel, the increase in use of nuclear power systems in support of space exploration programmes, and the placement on-orbit of global utilities. Furthermore, it is worth repeating that the risks posed by space activities both on-ground and on-orbit are very much of an international nature. The necessity of developing a harmonized framework of international rules which would govern, in particular, certification of ground and flight systems, personnel, activities, as well as the establishment of a means to manage space traffic to prevent interference with air traffic and on-orbit operations, therefore comes as a logical consequence.

It is also significant to note that, currently, there is little interest and utility in regulating the region of space beyond the geosynchronous orbits, whereas there are substantial and strategic interests (civil/commercial and military) in the region of space up to and including the geosynchronous orbits. It is therefore proposed that a new international regulatory framework should be established for the near space region for the purpose of achieving the goals listed below.

The establishment of an international civil/commercial space regulatory framework would have far reaching beneficial effects in removing obstacles to international space trade, as well as meeting the growing demand of military commands for transparent and accountable use of space in time of peace.

Finally, the misconception that military interests could run contrary to the establishment of international civil regulations for space should be dispelled. Military commands have a keen interest in space traffic management as well as in the control of space debris. They also have an overall interest in transparent communication as a way of preventing military incidents. Currently, the U.S. military is exploring areas of cooperation with Russia and China that could play a role in helping assure officials from each country's military that the others do not have ulterior motives behind their use of space surveillance sensors. The countries involved are even pondering the development of joint standards for station keeping so that any one of them is not unnecessarily alarmed by a sudden manoeuvre of the other's satellite that was intended solely as a station keeping manoeuvre and not as the beginning of an attack. In 2007, the *Stimson Centre,* a Washington-based think-tank acting in consultation with experts from several countries including Canada, France, Japan, Russia, and the United States, published a "*Model Code of Conduct for Space-Faring Nations*". The Model Code is reproduced in Appendix C to this Study. In many respects, the proposed code provides a blue print for a substantial portion of an international civil space convention. The Code reaffirmed "the crucial importance of outer space for global economic progress, commercial advancement, scientific research, sustainable development, as well as national, regional and international security".

In view of the ever increasing number of international actors and space activities as well as serious risks, it is necessary to develop a harmonized framework of international rules which would include, in particular, uniform safety certification practices for ground and flight systems, personnel, as well as the establishment of the means to control space traffic to prevent interference with air traffic and on-orbit operations. It is therefore proposed that a new international regulatory framework should be established for the purpose of achieving the following goals:

i) Equally protect the citizens of all nations from the risks posed by launching, overflight, and re-entry of space systems;
ii) Develop, build, and operate space systems in accordance with common ground and flight safety rules, procedures and standards based on the status of knowledge and the accumulated experience of all space-faring nations;
iii) Establish international traffic control rules for launch, on-orbit, and re-entry operations to prevent collisions or interference with other space systems and with air traffic;
iv) Protect the ground, air, and space environments from chemical, radioactive, and debris contamination as a result of space operations;
v) Ban intentional destruction of any on-orbit space system or other harmful activities that pose safety and environmental risks; and
vi) Establish mutual aid provisions for space mission emergencies.

It is believed that an international regulatory framework is not only needed to protect citizens of all nations. It would also serve the purpose of meeting growing demands made by military commands for the transparent and accountable use of space by civil and commercial operators, ease the barriers to international space commerce that different national safety regulations may create, prevent distortion of commercial competition due to substandard safety practices, and allow for mutual assistance and rescue in case of need.

5.2.2 Regulatory model

It is apparent from the discussion above that ICAO is a fully operational legislative and implementing body ideally suited for taking up the issues identified herein in relation to space activities. Since ICAO has established detailed rules, regulations, guidelines, and operational procedures for aviation, it would be an easy and suitable job for ICAO to extend the same to space by appropriately amending its existing Annexes and/or adopting a few extra Annexes thereby extending its coverage to issues like licensing of spaceports, human space flight, space traffic management, safety of astronauts, security, etc. As suborbital flights are poised to become routine in the near future, now is an opportune time to establish regulations and standardization in connection therewith through ICAO.

In order to implement the ICAO for Space proposal, three key areas must be addressed. The first is the development of a *space safety oversight operating model*. The operating model describes the concept of operations and how the regulatory regime would work. The second is the establishment of an organizational framework for the implementation of the operating model. The section below on the *ICAO for Space Organization* discusses how this new organization should be constituted. Finally, the *safety certification process* also needs to be considered, and this concerns the various processes a commercial space entity would have to follow in order to obtain the relevant safety certification(s) and permission(s) to operate.

5.2.3 Space safety oversight operating model

The proposed ICAO for Space regulatory regime should be designed as a comprehensive, permanent, and holistic space safety oversight regulatory regime. Through international deliberations, ICAO for Space would set the overall space safety regulatory regime that national space authorities would implement. The

Tab. 5.1: *Key elements of international space safety regulatory regime*

Key element	Application
Independence	• The regulatory regime should be sufficiently independent from the Industry so that: • There are sufficient oversight arrangements to monitor activities • Involvement in the industry is limited to avoid jeopardizing independence
Clear Fundamentals	• ICAO should set clear safety fundamentals that national space authorities must meet • Safety fundamentals should be set in consultation with industry • This accomplishes joint ownership from industry and government
Dynamic	• Regulatory regime should be a dynamic process • Must be a risk-based process • Use this to set annual safety performance goals linked to national space authorities and certificate holders
Use Modern Tools	• Safety analysis must use modern system safety engineering tools • Focus on entire system – not just workplace safety • National space authority safety regulator periodically review and audit to verify system safety tools being applied appropriately
Accountability	• Accountability must be tied to individuals • National space authority safety regulator must hold commercial space entity responsible and accountable for safety • Commercial space entity must indicate how management and staff are held accountable and verified by regulator
Proof of Safety Certification	• A proof of safety certification process should be the principal regulatory compliance vehicle • General proof of safety for initial certification followed by triennial reviews • Proof of safety updated made prior to material changes (management, operations, and infrastructure) • National space authority safety regulator should conduct periodic audits to verify compliance

national space authority as safety regulator would be responsible for carrying out safety certification of all commercial space activities in accordance with standards prescribed by the ICAO for Space organization. Table 5.1 illustrates the key elements and applications of such a regulatory regime.

ICAO for Space should focus on ensuring that commercial space entities base their management and assurance processes on a robust safety management system (SMS). In essence, the entire safety certification process must be predicated on evaluation, certification, and monitoring the SMS of commercial space entities. In addition, this meets ICAO's stated goal of integrating SMS into air traffic management and air flight standards.

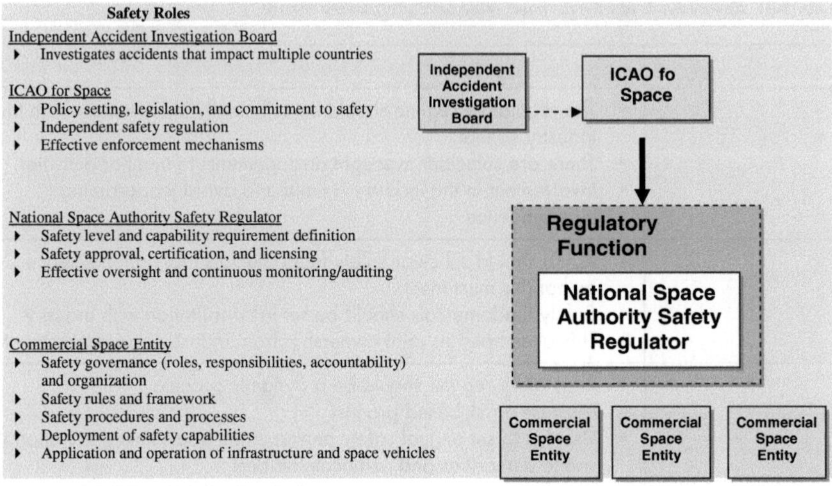

Fig. 5.1. *Proposed structure and safety roles.*

The commercial space entity's SMS should provide a systematic method to control risks and to provide assurance that those risk controls are effective. The commercial entity's SMS and its compliance with technical space safety regulations will be certified by the national space authority (the safety regulator) through an ICAO-sanctioned safety certification programme. The key concepts of the proposed ICAO for Space and the current ICAO aviation flight safety standards should be the same and are:

- Identification of safety hazards
- Ensuring that remedial action necessary to maintain an acceptable level of safety is implemented
- Providing for continuous monitoring and regular assessment of the safety level achieved
- Aiming to make continuous improvements to the overall level of safety (proposed amendment to ICAO Annex 6 for air flight standards)
- Clearly defining lines of safety accountability throughout the operator's organization, including direct accountability for safety on the part of senior management.

Figure 5.1 illustrates the proposed structure and safety roles for ICAO for Space, the national space authority safety regulator, and the commercial space entity.

Safety regulations encompass four critical functions: policy and regulation setting, compliance enforcement, continuous monitoring, and independent accident investigation. The key activities are:

Policy and regulation setting – This includes the development and periodic review of space safety regulations. It should address key issues for rulemaking and ensure that stakeholder (i.e., the commercial space community) outreach is part of the rulemaking process. As pertains in the current aviation safety programme at ICAO, general safety guidelines and rules as well as specific technical safety requirements should be developed at the international level. The national space authority safety regulator would develop its own set of regulations and oversight processes that meet ICAO requirements and yet still maintain a national focus. The national space authority safety regulator would be responsible for ensuring that ICAO and national regulations are implemented at the domestic level.

Compliance enforcement – ICAO would set guidelines for the compliance enforcement regime to be implemented by the national space authority safety regulator. It would be the responsibility of the national space authority safety regulator to conduct safety oversight of the commercial space entities operating within its national boundaries. The national space authority safety regulator would review and approve the commercial entity's safety certification process.

Monitoring – Regular audits and inspections are the central actions of the monitoring process. Though ICAO would set the monitoring guidelines and regimes, actual monitoring activities would be conducted by the national space authority safety regulator. In addition, safety trend monitoring and incident and accident investigations would be carried out by the national space authority safety regulator and ICAO.

Independent accident investigation – Because space activities and potential accidents most likely will cross international borders, it is imperative to have an independent accident investigation board established within ICAO. The investigation board's remit would be strictly investigation and reporting, and it would have no compliance or regulatory oversight functions. The intent is for the board to have a formal process for understanding what went wrong and communicating the lessons learned therefrom to the international commercial space community.

5.2.4 ICAO for space organization

Figure 5.2 illustrates the current organizational structure of ICAO. As can be seen, the organization is primarily focused on aviation safety and navigation.

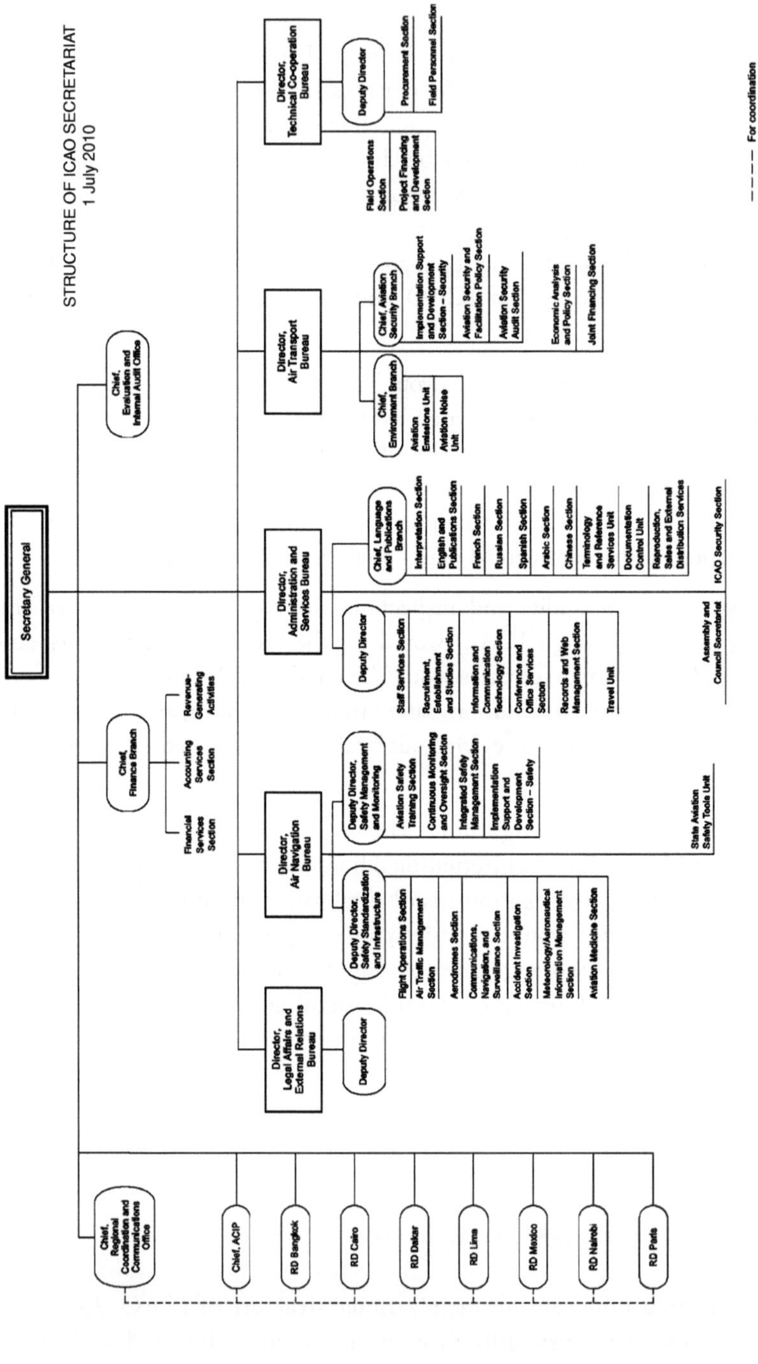

Fig. 5.2. *Current ICAO organizational structure.*

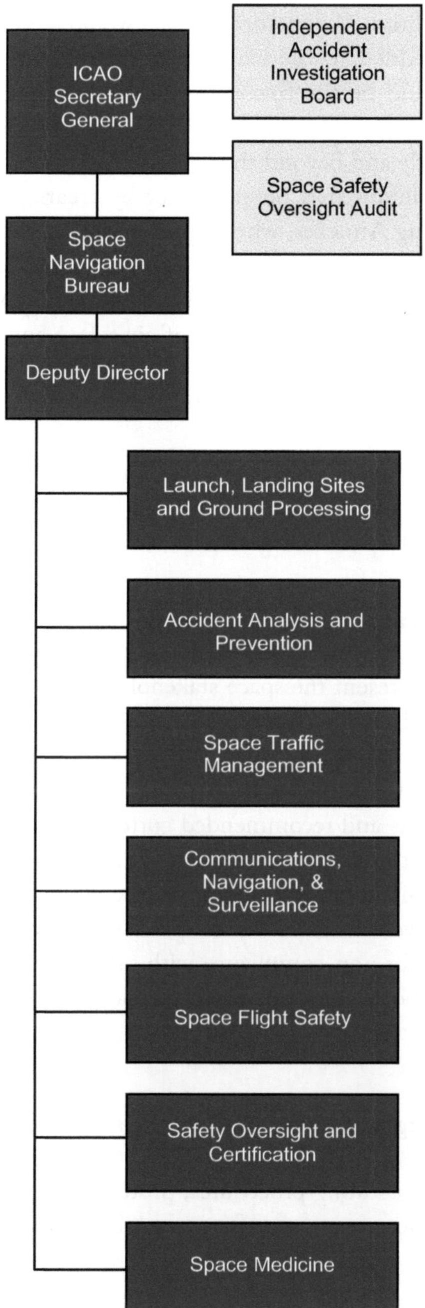

Fig. 5.3. *Proposed ICAO for space organization chart.*

Eventually, an additional directorate that would report to the Secretary General might be added to the existing structure. A possible name for the new Directorate could be the *Space Navigation Bureau.* Also, a space safety oversight audit section could be added to the Office of the Secretary General. But more importantly and beyond the organizational structures, ICAO should address the issues illuminated herein either by creating a new Annex or by amending the existing Annexes, where appropriate, to address aerospace safety and navigation issues. Ample jurisdiction for such a reform exists under Article 37 of the Chicago Convention, which confers upon ICAO, the authority to address not only the issues of air safety and navigation, but also to address, and "such other matters concerned with the safety, regularity, and efficiency of air navigation as may from time to time appear appropriate". That time has come.

The purpose of the Space Navigation Bureau might be to ensure a viable, long-term, and continuous oversight process. Figure 5.3 illustrates the major components of the proposed Space Navigation Bureau.

Independent Accident Investigation Board – reports to the Secretary General

- Board members represent the space stakeholder community
- Independent board charged with investigating all major commercial space-based accidents
- Full investigation authority (but non-regulatory)
- Investigation reports and recommended corrective actions are non-regulatory
- Body is non-regulatory so that all the facts of an investigation can be brought to light in a non-punitive fashion. However, the national space authority safety regulator does have the authority to penalize individuals, companies, and operating entities for non-compliance with safety regulations
- The investigation board does not apportion blame or liability nor enforce law or carry out prosecutions

Space Safety Oversight Audit Unit – reports to the Secretary General

- Develops standard auditing procedures, protocols, and other tools
- Plans, conducts and reports on safety oversight audits and audit follow ups
- Develops guidance material on safety oversight-related subjects

Space Navigation Bureau – develops studies and proposes Standards and Recommended Practices (SARPs) relating to the safety of space navigation.

Within the Space Navigation Bureau, the following units may be created:

Launch/Landing Sites and Ground Processing – responsible for launch, landing, and ground processing activities:

- Launch/landing site infrastructure planning and design
- Site safety certification
- Site services, including rescue and fire fighting, site emergency preparedness and planning, and site maintenance
- Launch/landing site operational services

Accident Analysis and Prevention – analyzes accident, incidents, and near miss trends:

- Develops investigation SARPs for implementation by the national space authority
- Monitors developments in accident investigation and prevention techniques and practices
- Provides guidance on the collection and analysis of accident and incident data
- Monitors developments in safety management systems concepts and practices
- Analyzes and trends accident, incident, and near miss data

Space Traffic Management – is responsible for the concept of operations of space traffic management worldwide. It also develops and promulgates SARPs to the national space authority safety regulator:

- Evaluates performance of space traffic management in terms of: safety, regularity, efficiency, safety certification, and quality assurance
- Addresses key traffic management safety issues such as:
    - Autonomy of flight
    - Situational awareness
    - Separation assurance
    - Collision avoidance
    - Optimization of traffic flows
    - General space traffic management
- Develops STM requirements for communications, navigation, and surveillance as related to orbital and suborbital flights
- Space traffic flow management of orbital and suborbital flights

Communication, Navigation, and Surveillance – evaluates, coordinates, and tracks communication and navigation issues:

- Provides studies and guidance on communications, radio navigation, and surveillance concepts and functions
- Coordinates with the ITU on all matters concerning the radio frequency spectrum allocated to space communications, navigation, and surveillance services

Space Flight Safety – responsible for the development of SARPs and guidance material related to the operation, certification, and space-worthiness (all phases from launch to re-entry/disposal) of space craft including design, licensing and training of personnel and crew, and the safe transport of dangerous goods through air and space (e.g., radioactive materials).

Safety Oversight and Certification – sets the overall safety oversight and certification process, guidelines and SARPs for the national space authority safety regulator to implement

- Manages the safety certification process
- Responsible for the harmonization of decision-making criteria for the processes for safety certification and safety licensing by the national space authority safety regulator
- Provides guidance on how commercial space entities can qualify for safety certification
- A new safety certification must be performed for all major infrastructure, hardware, software, operational, or management changes
- Re-certification must be submitted at least every five years

Space Medicine – sets the medical requirements for space craft crew members and passengers:

- Monitors the developments within the field of space medicine
- Provides guidance to national space authority safety regulators on medical standards, medical problems, effect of working conditions on health, and the biological and psychological problems related to space passengers and crew, first aid, and survival equipment

5.2.5 General description of the safety certification process

Safety certification is a formal mechanism for ensuring that the commercial space entity meets its safety obligations. The national space authority safety regulator reviews and approves requests from commercial space entities for safety certification. The ICAO for Space organization sets the safety certification requirements. The purpose of safety certification is to ensure that:

- the commercial space entity meets pre-defined safety standards and norms
- the certification process is premised on certifying the commercial space entity's safety management system and its compliance with technical regulations promulgated by ICAO and the national space authority safety regulator
- all commercial space entities are regularly certified by a national space authority safety regulator or else they cannot operate

Tab. 5.2: *Safety certification process*

• To become certified, a commercial space entity must submit a completed application form, a safety certification package, and supporting information to the national space authority safety regulator
• The safety certification package must include a detailed description and evidence of the commercial entity's safety management system (SMS)
• Additional supporting technical information includes:
◦ Vehicle and system descriptions
◦ Operational environment
◦ Pre/post flight operations
◦ Operating area containment
◦ Flight rules
◦ Tracking and communications
◦ Etc.
• The national space authority safety regulator will review the application
◦ Conduct an audit to determine whether the applicant's SMS and technical supporting documentation satisfies the relevant SARPs
◦ Determine whether the applicant has the competency and capacity to implement and comply with their SMS
• Commercial space entities interested in applying for an exemption from some of the safety certification requirements should contact the national space authority safety regulator before applying to determine if they are eligible for an exemption.

- for a commercial space entity operation to be safety-approved, national authorities must be satisfied that
  - the commercial space entity has the competency and capacity to implement a robust safety management system (SMS)
  - the SMS effectively controls the risks associated with carrying out its launch-to-disposal operations life cycle
- the national space authority safety regulator conducts regular audits and inspections of operations
- failure to comply with legal requirements and safety certification conditions will result in certification being varied, suspended, or revoked
- commercial space entities conduct their own internal safety certification which forms the basis of an application to the national space authority safety regulator for safety certification.

Table 5.2 lists the primary steps envisaged in the safety certification process. Figure 5.4 illustrates the safety certification review and approval process and Exhibit 8-7 shows the kind of internal commercial space entity data flows that would be required to support a safety certification process.

5.2.6 Suggested ICAO for space regulatory implementation

Figure 5.4 below illustrates a suggested process to implement a set of ICAO space standards and recommended practices (SARPs).

In order to implement the above-suggested ICAO for Space regulatory steps and mechanisms, it may be appropriate to follow an orderly and systematic approach. Undoubtedly, the ideal solution to accommodate space traffic management and other space safety requirements would be to amend the Chicago Convention thereby expressly extending ICAO's jurisdiction over space. However, it is well-known that bringing an amendment to the Chicago Convention into force takes approximately 25 years to accomplish. Thus, there is the need to explore alternatives. There are three possibilities.

One is the approach based on the residual powers of ICAO as set forth in Article 37 of the Chicago Convention giving it jurisdiction over everything that has an impact on the safety, regularity and efficiency of international air navigation. Since space traffic has such an impact on the safety of civil aviation, it is legitimate for

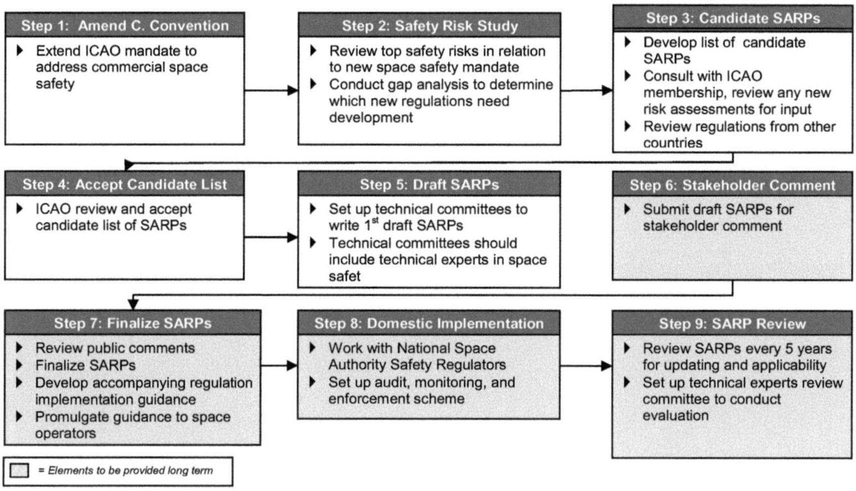

Fig. 5.4. *Suggested ICAO for space SARPs development process.*

ICAO to regulate or coordinate space traffic and such other directly related space matters. Over time, ICAO may by customary evolution acquire competence in regulating space traffic and other space related matters. The second is that the Council may make appropriate changes to the ICAO Annexes, thereby expanding ICAO's jurisdiction over space matters, starting at the very least with those space activities that have an impact of on civil aviation. The third approach is to adopt a new international treaty, to be ratified by a small number of States but including all States whose interests are particularly affected. This treaty may specifically entrust ICAO with the function of conducting space traffic management and other safety regulation. In line with the first approach, ICAO could, in the interim, allow the current Air Navigation Bureau to gradually expand the scope of its activities to the space sector, instead of establishing a full-fledged space navigation bureau. When the latter is established in the future, respective jurisdictional conflicts between the resulting air navigation and space navigation bureaux ought to be avoided.

Appendix A
Relevant excerpts of the ITU constitution and convention

(as amended by the Plenipotentiary Conferences in 2006,[1] and in 2010)

Constitution of the International Telecommunication Union

Preamble

1 While fully recognizing the sovereign right of each State to regulate its telecommunication and having regard to the growing importance of telecommunication for the preservation of peace and the economic and social development of all States, the States Parties to this Constitution, as the basic instrument of the International Telecommunication Union, and to the Convention of the International Telecommunication Union (hereinafter referred to as "the Convention") which complements it, with the object of facilitating peaceful relations, international cooperation among peoples and economic and social development by means of efficient telecommunication services, have agreed as follows.

Article 1: Purposes of the union

2 1 The purposes of the Union are:

3 *a)* to maintain and extend international cooperation among all its Member States for the improvement and rational use of telecommunications of all kinds;

3A *abis)* to promote and enhance participation of entities and organizations in the activities of the Union and foster fruitful cooperation and partnership between them and Member States for the fulfilment of the overall objectives as embodied in the purposes of the Union;

4 *b)* to promote and to offer technical assistance to developing countries in the field of telecommunications, and also to promote the mobilization of the material, human and financial resources needed for its implementation, as well as access to information;

5 *c)* to promote the development of technical facilities and their most efficient operation with a view to improving the efficiency of telecommunication services, increasing their usefulness and making them, so far as possible, generally available to the public;

6 *d*) to promote the extension of the benefits of the new telecommunication technologies to all the world's inhabitants;

7 *e*) to promote the use of telecommunication services with the objective of facilitating peaceful relations;

8 *f*) to harmonize the actions of Member States and promote fruitful and constructive cooperation and partnership between Member States and Sector Members in the attainment of those ends;

9 *g*) to promote, at the international level, the adoption of a broader approach to the issues of telecommunications in the global information economy and society, by cooperating with other world and regional intergovernmental organizations and those non-governmental organizations concerned with telecommunications.

10 2 To this end, the Union shall in particular:

11 *a*) effect allocation of bands of the radio-frequency spectrum, the allotment of radio frequencies and the registration of radio-frequency assignments and, for space services, of any associated orbital position in the geostationary-satellite orbit or of any associated characteristics of satellites in other orbits, in order to avoid harmful interference between radio stations of different countries;

12 *b*) coordinate efforts to eliminate harmful interference between radio stations of different countries and to improve the use made of the radio-frequency spectrum for radiocommunication services and of the geostationary-satellite and other satellite orbits;

13 *c*) facilitate the worldwide standardization of telecommunications, with a satisfactory quality of service;

14 *d*) foster international cooperation and solidarity in the delivery of technical assistance to the developing countries and the creation, development and improvement of telecommunication equipment and networks in developing countries by every means at its disposal, including through its participation in the relevant programmes of the United Nations and the use of its own resources, as appropriate;

15 *e*) coordinate efforts to harmonize the development of telecommunication facilities, notably those using space techniques, with a view to full advantage being taken of their possibilities;

16 *f*) foster collaboration among Member States and Sector Members with a view to the establishment of rates at levels as low as possible consistent with an efficient service and taking into account the necessity for maintaining independent financial administration of telecommunications on a sound basis;

17 *g)* promote the adoption of measures for ensuring the safety of life through the cooperation of telecommunication services;

18 *h)* undertake studies, make regulations, adopt resolutions, formulate recommendations and opinions, and collect and publish information concerning telecommunication matters;

19 *i)* promote, with international financial and development organizations, the establishment of preferential and favourable lines of credit to be used for the development of social projects aimed, *inter alia*, at extending telecommunication services to the most isolated areas in countries;

19A *j)* promote participation of concerned entities in the activities of the Union and cooperation with regional and other organizations for the fulfilment of the purposes of the Union.

Article 4: Instruments of the union

29 1 The instruments of the Union are:

- this Constitution of the International Telecommunication Union,
- the Convention of the International Telecommunication Union and
- the Administrative Regulations.

30 2 This Constitution, the provisions of which are complemented by those of the Convention, is the basic instrument of the Union.

31 3 The provisions of both this Constitution and the Convention are further complemented by those of the Administrative Regulations, enumerated below, which regulate the use of telecommunications and shall be binding on all Member States:

- International Telecommunication Regulations,
- Radio Regulations.

32 4 In the case of inconsistency between a provision of this Constitution and a provision of the Convention or of the Administrative Regulations, the Constitution shall prevail. In the case of inconsistency between a provision of the Convention and a provision of the Administrative Regulations, the Convention shall prevail.

Article 6: Execution of the instruments of the union

37 1 The Member States are bound to abide by the provisions of this Constitution, the Convention and the Administrative Regulations in all telecommunication offices and stations established or operated by them which engage in international services or which are capable of causing harmful interference to radio services of other countries, except in regard to services exempted from these obligations in accordance with the provisions of Article 48 of this Constitution.

38 2 The Member States are also bound to take the necessary steps to impose the observance of the provisions of this Constitution, the Convention and the Administrative Regulations upon operating agencies authorized by them to establish and operate telecommunications and which engage in international services or which operate stations capable of causing harmful interference to the radio services of other countries.

Article 7: Structure of the union

39 The Union shall comprise:

40 *a*) the Plenipotentiary Conference, which is the supreme organ of the Union;

41 *b*) the Council, which acts on behalf of the Plenipotentiary Conference;

42 *c*) world conferences on international telecommunications;

43 *d*) the Radiocommunication Sector, including world and regional radiocommunication conferences, radiocommunication assemblies and the Radio Regulations Board;

44 *e*) the Telecommunication Standardization Sector, including world telecommunication standardization assemblies;

45 *f*) the Telecommunication Development Sector, including world and regional telecommunication development conferences;

46 *g*) the General Secretariat.

Chapter II : Radiocommunication sector

Article 12 : Functions and structure

78 1 1) The functions of the Radiocommunication Sector shall be, bearing in mind the particular concerns of developing countries, to fulfil the purposes of the Union, as stated in Article 1 of this Constitution, relating to radiocommunication:

- by ensuring the rational, equitable, efficient and economical use of the radio-frequency spectrum by all radiocommunication services, including those using the geostationary-satellite or other satellite orbits, subject to the provisions of Article 44 of this Constitution and
- by carrying out studies without limit of frequency range and adopting recommendations on radiocommunication matters.

79 2) The precise responsibilities of the Radiocommunication Sector and the Telecommunication Standardization Sector shall be subject to continuing review, in close cooperation, with regard to matters of common interest to both Sectors, in accordance with the relevant provisions of the Convention. Close coordination shall be carried out between the Radiocommunication, Telecommunication Standardization and Telecommunication Development Sectors.

80 2 The Radiocommunication Sector shall work through:

81 *a*) world and regional radiocommunication conferences;

82 *b*) the Radio Regulations Board;

83 *c*) radiocommunication assemblies;

84 *d*) radiocommunication study groups;

84A *dbis*) the radiocommunication advisory group;

85 *e*) the Radiocommunication Bureau, headed by the elected Director.

86 3 The Radiocommunication Sector shall have as members:

87 *a*) of right, the administrations of all Member States;

88 *b*) any entity or organization which becomes a Sector Member in accordance with the relevant provisions of the Convention.

Chapter III: Telecommunication standardization sector

Article 17: Functions and structure

104 1 1) The functions of the Telecommunication Standardization Sector shall be, bearing in mind the particular concerns of the developing countries, to fulfil the purposes of the Union relating to telecommunication standardization, as stated in Article 1 of this Constitution, by studying technical, operating and tariff questions and adopting recommendations on them with a view to standardizing telecommunications on a worldwide basis.

105 2) The precise responsibilities of the Telecommunication Standardization and Radiocommunication Sectors shall be subject to continuing review, in close cooperation, with regard to matters of common interest to both Sectors, in accordance with the relevant provisions of the Convention. Close coordination shall be carried out between the Radiocommunication, Telecommunication Standardization and Telecommunication Development Sectors.

106 2 The Telecommunication standardization sector shall work through:

107 *a*) world telecommunication standardization assemblies;

108 *b*) telecommunication standardization study groups;

108A *bbis*) the telecommunication standardization advisory group;

109 *c*) the Telecommunication Standardization Bureau headed by the elected Director.

110 3 The Telecommunication Standardization Sector shall have as members:

111 *a*) of right, the administrations of all Member States;

112 *b*) any entity or organization which becomes a Sector Member in accordance with the relevant provisions of the Convention.

Article 18: World telecommunication standardization assemblies

113 1 The duties of world telecommunication standardization assemblies are specified in the Convention.

114 2 World telecommunication standardization assemblies shall be convened every four years; however, an additional assembly may be held in accordance with the relevant provisions of the Convention.

115 3 Decisions of world telecommunication standardization assemblies must in all circumstances be in conformity with this Constitution, the Convention and the Administrative Regulations. When adopting resolutions and decisions, the assemblies shall take into account the foreseeable financial implications and should avoid adopting resolutions and decisions which might give rise to expenditure in excess of the financial limits laid down by the Plenipotentiary Conference.

Article 39: Notification of infringements

190 In order to facilitate the application of the provisions of Article 6 of this Constitution, Member States undertake to inform and, as appropriate, assist one another with regard to infringements of the provisions of this Constitution, of the Convention and of the Administrative Regulations.

Chapter VII: Special provisions for radio

Article 44: Use of the Radio-Frequency spectrum and of the Geostationary-Satellite and other satellite orbits

195 1 Member States shall endeavour to limit the number of frequencies and the spectrum used to the minimum essential to provide in a satisfactory manner the necessary services. To that end, they shall endeavour to apply the latest technical advances as soon as possible.

196 2 In using frequency bands for radio services, Member States shall bear in mind that radio frequencies and any associated orbits, including the geostationary-satellite orbit, are limited natural resources and that they must be used rationally, efficiently and economically, in conformity with the provisions of the Radio Regulations, so that countries or groups of countries may have equitable access to those orbits and frequencies, taking into account the special needs of the developing countries and the geographical situation of particular countries.

Article 45: Harmful interference

197 1 All stations, whatever their purpose, must be established and operated in such a manner as not to cause harmful interference to the radio services or communications of other Member States or of recognized operating agencies, or of other duly authorized operating agencies which carry on a radio service, and which operate in accordance with the provisions of the Radio Regulations.

198 2 Each Member State undertakes to require the operating agencies which it recognizes and the other operating agencies duly authorized for this purpose to observe the provisions of No. 197 above.

199 3 Further, the Member States recognize the necessity of taking all practicable steps to prevent the operation of electrical apparatus and installations of all kinds from causing harmful interference to the radio services or communications mentioned in No. 197 above.

Article 46: Distress calls and messages

200 Radio stations shall be obliged to accept, with absolute priority, distress calls and messages regardless of their origin, to reply in the same manner to such messages, and immediately to take such action in regard thereto as may be required.

Article 47: False or deceptive distress, urgency, safety or identification signals

201 Member States agree to take the steps required to prevent the transmission or circulation of false or deceptive distress, urgency, safety or identification signals, and to collaborate in locating and identifying stations under their jurisdiction transmitting such signals.

Article 48: Installations for national defence services

202 1 Member States retain their entire freedom with regard to military radio installations.

203 2 Nevertheless, these installations must, so far as possible, observe statutory provisions relative to giving assistance in case of distress and to the measures to be taken to prevent harmful interference, and the provisions of the Administrative

Regulations concerning the types of emission and the frequencies to be used, according to the nature of the service performed by such installations.

204 3 Moreover, when these installations take part in the service of public correspondence or other services governed by the Administrative Regulations, they must, in general, comply with the regulatory provisions for the conduct of such services.

Convention of the International Telecommunication Union

Section 6: Telecommunication Standardization Sector

Article 13: World Telecommunication Standardization Assembly

184 1 In accordance with No. 104 of the Constitution, a world telecommunication standardization assembly shall be convened to consider specific matters related to telecommunication standardization.

184A 1*bis*) The world telecommunication standardization assembly is authorized to adopt the working methods and procedures for the management of the Sector's activities in accordance with No. 145A of the Constitution.

185 2 The questions to be studied by a world telecommunication standardization assembly, on which recommendations shall be issued, shall be those adopted pursuant to its own procedures or referred to it by the Plenipotentiary Conference, any other conference, or the Council.

186 3 In accordance with No. 104 of the Constitution, the assembly shall:

187 *a*) consider the reports of study groups prepared in accordance with No. 194 of this Convention and approve, modify or reject draft recommendations contained in those reports, and consider the reports of the telecommunication standardization advisory group in accordance with Nos. 197H and 197I of this Convention;

188 *b*) bearing in mind the need to keep the demands on the resources of the Union to a minimum, approve the programme of work arising from the review of existing questions and new questions and determine the priority, urgency, estimated financial implications and time-scale for the completion of their study;

189 *c*) decide, in the light of the approved programme of work derived from No. 188 above, on the need to maintain, terminate or establish study groups and allocate to each of them the questions to be studied;

190 *d*) group, as far as practicable, questions of interest to the developing countries to facilitate their participation in these studies;

191 *e)* consider and approve the report of the Director on the activities of the Sector since the last conference;

191A *f)* decide on the need to maintain, terminate or establish other groups and appoint their Chairmen and Vice-Chairmen;

191B *g)* establish the terms of reference for the groups referred to in No. 191A above; such groups shall not adopt questions or recommendations.

191C 4 A world telecommunication standardization assembly may assign specific matters within its competence to the telecommunication standardization advisory group indicating the action required on those matters.

191D 5 A world telecommunication standardization assembly shall be presided over by a Chairman designated by the government of the country in which the meeting is held or, in the case of a meeting held at the seat of the Union, by a Chairman elected by the assembly itself. The Chairman shall be assisted by Vice-Chairmen elected by the assembly.

Article 14: Telecommunication standardization study groups

192 1 1) Telecommunication standardization study groups shall study questions adopted in accordance with a procedure established by the world telecommunication standardization assembly and prepare draft recommendations to be adopted in accordance with the procedure set forth in Nos. 246A–247 of this Convention.

193 2) The study groups shall, subject to No. 195 below, study technical, operating and tariff questions and prepare recommendations on them with a view to standardizing telecommunications on a worldwide basis, including recommendations on interconnection of radio systems in public telecommunication networks and on the performance required for these interconnections. Technical or operating questions specifically related to radiocommunication as enumerated in Nos. 151–154 of this Convention shall be within the purview of the Radiocommunication Sector.

194 3) Each study group shall prepare for the world telecommunication standardization assembly a report indicating the progress of work, the recommendations adopted in accordance with the consultation procedure contained in No. 192 above, and any draft new or revised recommendations for consideration by the assembly.

195 2 Taking into account No. 105 of the Constitution, the tasks enumerated in No. 193 above and those enumerated in Nos. 151–154 of this Convention in relation to the Radiocommunication Sector shall be kept under continuing review by the Telecommunication Standardization Sector and the Radiocommunication Sector with a view to reaching common agreement on changes in the distribution of matters under study. The two Sectors shall cooperate closely and adopt procedures to conduct such a review and reach agreements in a timely and effective manner. If agreement is not reached, the matter may be submitted through the Council to the Plenipotentiary Conference for decision.

196 3 In the performance of their studies, the telecommunication standardization study groups shall pay due attention to the study of questions and to the formulation of recommendations directly connected with the establishment, development and improvement of telecommunications in developing countries at both the regional and international levels. They shall conduct their work giving due consideration to the work of national, regional and other international standardization organizations, and cooperate with them, keeping in mind the need for the Union to maintain its pre-eminent position in the field of worldwide standardization for telecommunications.

197 4 For the purpose of facilitating the review of activities in the Telecommunication Standardization Sector, measures should be taken to foster cooperation and coordination with other organizations concerned with telecommunication standardization and with the Radiocommunication Sector and the Telecommunication Development Sector. A world telecommunication standardization assembly shall determine the specific duties, conditions of participation and rules of procedure for these measures.

Article 14A: Telecommunication standardization advisory group

197A 1 The telecommunication standardization advisory group shall be open to representatives of administrations of Member States and representatives of Sector Members and to chairmen of the study groups and other groups.

197B 2 The telecommunication standardization advisory group shall:

197C 1) review priorities, programmes, operations, financial matters and strategies for activities in the Telecommunication Standardization Sector;

197CA 1*bis*) review the implementation of the operational plan of the preceding period in order to identify areas in which the Bureau has not achieved or was not able to achieve the objectives laid down in that plan, and advise the Director on the necessary corrective measures;

197D 2) review progress in the implementation of the programme of work established under No. 188 of this Convention;

197E 3) provide guidelines for the work of study groups;

197F 4) recommend measures, *inter alia*, to foster cooperation and coordination with other relevant bodies, with the Radiocommunication Sector, the Telecommunication Development Sector and the General Secretariat;

197G 5) adopt its own working procedures compatible with those adopted by the world telecommunication standardization assembly;

197H 6) prepare a report for the Director of the Telecommunication Standardization Bureau indicating action in respect of the above items.

197I 7) prepare a report for the world telecommunication standardization assembly on the matters assigned to it in accordance with No. 191A and transmit it to the Director for submission to the assembly.

Article 15: Telecommunication standardization bureau

198 1 The Director of the Telecommunication Standardization Bureau shall organize and coordinate the work of the Telecommunication Standardization Sector.

199 2 The Director shall, in particular:

200 *a*) update annually the work programme approved by the world telecommunication standardization assembly, in consultation with the chairmen of the telecommunication standardization study groups and other groups;

201 *b*) participate, as of right, but in an advisory capacity, in the deliberations of world telecommunication standardization assemblies and of the telecommunication standardization study groups and other groups. The Director shall make all necessary preparations for assemblies and meetings of the Telecommunication Standardization Sector in consultation with the General Secretariat in accordance with No. 94 of this Convention and, as appropriate, with the other Sectors of the Union, and with due regard for the directives of the Council concerning these preparations;

202 *c*) process information received from administrations in application of the relevant provisions of the International Telecommunication Regulations or decisions of the world telecommunication standardization assembly and prepare it, where appropriate, in a suitable form for publication;

203 *d*) exchange with Member States and Sector Members data in machine-readable and other forms, prepare and, as necessary, keep up to date any documents and databases of the Telecommunication Standardization Sector, and arrange with the Secretary-General, as appropriate, for their publication in the languages of the Union in accordance with No. 172 of the Constitution;

204 *e*) submit to the world telecommunication standardization assembly a report on the activities of the Sector since the last assembly; the Director shall also submit to the Council and to the Member States and Sector Members such a report covering the two-year period since the last assembly, unless a second assembly is convened;

205 *f*) prepare a cost-based budget estimate for the requirements of the Telecommunication Standardization Sector and transmit it to the Secretary-General for consideration by the Coordination Committee and inclusion in the Union's budget;

205A *g*) prepare annually a rolling four-year operational plan that covers the subsequent year and the following three-year period, including financial implications of activities to be undertaken by the Bureau in support of the Sector as a whole; this four-year operational plan shall be reviewed by the telecommunication standardization advisory group in accordance with Article 14A of this Convention, and shall be reviewed and approved annually by the Council;

205B *h*) provide the necessary support for the telecommunication standardization advisory group, and report each year to Member States and Sector Members and to the Council on the results of its work;

205C *i*) provide assistance to developing countries in the preparatory work for world standardization assemblies, particularly with regard to matters of a priority nature for those countries.

206 3 The Director shall choose the technical and administrative personnel of the Telecommunication Standardization Bureau within the framework of the budget as approved by the Council. The appointment of the technical and administrative personnel is made by the Secretary-General in agreement with the Director. The final decision on appointment or dismissal rests with the Secretary-General.

207 4 The Director shall provide technical support, as necessary, to the Telecommunication Development Sector within the framework of the Constitution and this Convention.

Article 20: Conduct of business of study groups

242 1 The radiocommunication assembly, the world telecommunication standardization assembly and the world telecommunication development conference shall appoint the chairman and one Vice-Chairman or more for each study group. In appointing Chairmen and Vice-Chairmen, particular consideration shall be given to the requirements of competence and equitable geographical distribution, and to the need to promote more efficient participation by the developing countries.

243 2 If the workload of any study group requires, the assembly or conference shall appoint such additional Vice-Chairmen as it deems necessary.

244 3 If, in the interval between two assemblies or conferences of the Sector concerned, a study group Chairman is unable to carry out his duties and only one Vice-Chairman has been appointed, then that Vice-Chairman shall take the Chairman's place. In the case of a study group for which more than one Vice-

Chairman has been appointed, the study group at its next meeting shall elect a new Chairman from among those Vice-Chairmen and, if necessary, a new Vice-Chairman from among the members of the study group. It shall likewise elect a new Vice-Chairman if one of the Vice-Chairmen is unable to carry out his duties during that period.

245 4 Study groups shall conduct their work as far as possible by correspondence, using modern means of communication.

246 5 The Director of the Bureau of each Sector, on the basis of the decisions of the competent conference or assembly, after consultation with the Secretary-General and coordination as required by the Constitution and Convention, shall draw up the general plan of study group meetings.

246A 5*bis*) 1) Member States and Sector Members shall adopt questions to be studied in accordance with procedures established by the relevant conference or assembly, as appropriate, including the indication whether or not a resulting recommendation shall be the subject of a formal consultation of Member States.

246B 2) Recommendations resulting from the study of the above questions are adopted by a study group in accordance with procedures established by the relevant conference or assembly, as appropriate. Those recommendations which do not require formal consultation of Member States for their approval shall be considered as approved.

246C 3) A recommendation requiring formal consultation of Member States shall be either treated in accordance with No. 247 below or transmitted to the relevant conference or assembly, as appropriate.

246D 4) Nos. 246A and 246B above shall not be used for questions and recommendations having policy or regulatory implications such as:

246E *a*) questions and recommendations approved by the Radiocommunication Sector relevant to the work of radiocommunication conferences, and other categories of questions and recommendations that may be decided by the radiocommunication assembly;

246F *b*) questions and recommendations approved by the Telecommunication Standardization Sector which relate to tariff and accounting issues, and relevant numbering and addressing plans;

246G *c*) questions and recommendations approved by the Telecommunication Development Sector which relate to regulatory, policy and financial issues;

246H *d*) questions and recommendations where there is any doubt about their scope.

247 6 Study groups may initiate action for obtaining approval from Member States for recommendations completed between two assemblies or conferences. The procedures to be applied for obtaining such approval shall be those approved by the competent assembly or conference, as appropriate.

247A 6*bis*) Recommendations approved in application of Nos. 246B or 247 above shall have the same status as ones approved by the conference or assembly itself.

248 7 Where necessary, joint working parties may be established for the study of questions requiring the participation of experts from several study groups.

248A 7*bis*) Following a procedure developed by the Sector concerned, the Director of a Bureau may, in consultation with the chairman of the study group concerned, invite an organization which does not participate in the Sector to send representatives to take part in the study of a specific matter in the study group concerned or its subordinate groups.

248B 7*ter*) An Associate, as referred to in No. 241A of this Convention, will be permitted to participate in the work of the selected study group without taking part in any decision-making or liaison activity of that study group.

249 8 The Director of the relevant Bureau shall send the final reports of the study groups to the administrations, organizations and entities participating in the Sector. Such reports shall include a list of the recommendations approved in conformity with No. 247 above. These reports shall be sent as soon as possible and, in any event, in time for them to be received at least one month before the date of the next session of the conference concerned.

Resolution 16 (Rev. Minneapolis, 1998)

Refinement of the Radiocommunication Sector and Telecommunication Standardization Sector

The Plenipotentiary Conference of the International Telecommunication Union (Minneapolis, 1998),

Noting

the report by the Council on the results of the implementation of Resolution 16 (Kyoto, 1994),

Considering

a) that ITU should be the pre-eminent global standardization body in the telecommunication field, including radiocommunication;

b) that ITU is the pre-eminent body for efficient worldwide cooperation in the radio regulatory field;

c) that the Additional Plenipotentiary Conference (Geneva, 1992) recognized Nos. 78 and 104 of the Constitution as an initial allocation of work between the Radiocommunication (ITU-R) and Telecommunication Standardization (ITU-T) Sectors and outlined general principles and guidelines pertaining to the allocation of work between ITU-R and ITU-T;

d) that, in application of instructions handed down by the Additional Plenipotentiary Conference (Geneva, 1992), the World Telecommunication Standardization Conference (Helsinki, 1993) and the Radiocommunication Assembly (Geneva, 1993) adopted resolutions that establish procedures for ongoing review and, as appropriate, allocation of work, in order to achieve goals in terms of effectiveness and efficiency;

e) the need to involve all interested participants of ITU-R and ITU-T in this ongoing review;

f) that, when implementing this resolution, questions that may have implications for the International Telecommunication Regulations and the Radio Regulations require a more cautious approach.

Resolves

1 that the current process, in conformity with the relevant resolutions of the world telecommunication standardization conference and the radiocommunication assembly which provide for ongoing review of new and existing work and its allocation to ITU-R and ITU-T, shall be maintained;

2 that changes in the allocation of work between ITU-R and ITU-T on matters that may be related to the International Telecommunication Regulations or the Radio Regulations shall not be considered within that process.

[1] Available online: https://www.itu.int/net/about/basic-texts/index.aspx (last accessed: 03 January 2011).

Appendix B

Relevant excerpts of the convention on international civil aviation (Signed at Chicago, on 7 December 1944) – Chicago convention

Excerpts from Part I: Air navigation

Chapter I: General principles and application of the convention

Article 1

Sovereignty

The contracting States recognize that every State has complete and exclusive sovereignty over the airspace above its territory.

Article 3

Civil and State aircraft

(a) This Convention shall be applicable only to civil aircraft, and shall not be applicable to State aircraft.
(b) Aircraft used in military, customs and police services shall be deemed to be State aircraft.
(c) No State aircraft of a contracting State shall fly over the territory of another State or land thereon without authorization by special agreement or otherwise, and in accordance with the terms thereof.
(d) The contracting States undertake, when issuing regulations for their State aircraft, that they will have due regard for the safety of navigation of civil aircraft.

Article 3*bis*

(a) The contracting States recognize that every State must refrain from resorting to the use of weapons against civil aircraft in flight and that, in case of interception, the lives of persons on board and the safety of aircraft must not be endangered. This provision shall not be interpreted as modifying in any way the rights and obligations of States set forth in the Charter of the United Nations.

(b) The contracting States recognize that every State, in the exercise of its sovereignty, is entitled to require the landing at some designated airport of a civil aircraft flying above its territory without authority or if there are reasonable grounds to conclude that it is being used for any purpose inconsistent with the aims of this Convention; it may also give such aircraft any other instructions to put an end to such violations. For this purpose, the contracting States may resort to any appropriate means consistent with relevant rules of international law, including the relevant provisions of this Convention, specifically paragraph (a) of this Article. Each contracting State agrees to publish its regulations in force regarding the interception of civil aircraft.

(c) Every civil aircraft shall comply with an order given in conformity with paragraph (b) of this Article. To this end each contracting State shall establish all necessary provisions in its national laws or regulations to make such compliance mandatory for any civil aircraft registered in that State or operated by an operator who has his principal place of business or permanent residence in that State. Each contracting State shall make any violation of such applicable laws or regulations punishable by severe penalties and shall submit the case to its competent authorities in accordance with its laws or regulations.

(d) Each contracting State shall take appropriate measures to prohibit the deliberate use of any civil aircraft registered in that State or operated by an operator who has his principal place of business or permanent residence in that State for any purpose inconsistent with the aims of this Convention. This provision shall not affect paragraph (a) or derogate from paragraphs (b) and (c) of this Article.

Article 11

Applicability of air regulations

Subject to the provisions of this Convention, the laws and regulations of a contracting State relating to the admission to or departure from its territory of aircraft engaged in international air navigation, or to the operation and navigation of such aircraft while within its territory, shall be applied to the aircraft of all contracting States without distinction as to nationality, and shall be complied with by such aircraft upon entering or departing from or while within the territory of that State.

Article 12

Rules of the air

Each contracting State undertakes to adopt measures to insure that every aircraft flying over or maneuvering within its territory and that every aircraft carrying its nationality mark, wherever such aircraft may be, shall comply with the rules and regulations relating to the flight and maneuver of aircraft there in force. Each contracting State undertakes to keep its own regulations in these respects uniform, to the greatest possible extent, with those established from time to time under this Convention. Over the high seas, the rules in force shall be those established under this Convention. Each contracting State undertakes to insure the prosecution of all persons violating the regulations applicable.

Article 25

Aircraft in distress

Each contracting State undertakes to provide such measures of assistance to aircraft in distress in its territory as it may find practicable, and to permit, subject to control by its own authorities, the owners of the aircraft or authorities of the State in which the aircraft is registered to provide such measures of assistance as may be necessitated by the circumstances. Each contracting State, when undertaking search for missing aircraft, will collaborate in coordinated measures which may be recommended from time to time pursuant to this Convention.

Article 26

Investigation of accidents

In the event of an accident to an aircraft of a contracting State occurring in the territory of another contracting State, and involving death or serious injury, or indicating serious technical defect in the aircraft or air navigation facilities, the State in which the accident occurs will institute an inquiry into the circumstances of the accident, in accordance, so far as its laws permit, with the procedure which may be recommended by the International Civil Aviation Organization. The State in which the aircraft is registered shall be given the opportunity to appoint observers to be present at the inquiry and the State holding the inquiry shall communicate the report and findings in the matter to that State.

Article 31

Certificates of airworthiness

Every aircraft engaged in international navigation shall be provided with a certificate of airworthiness issued or rendered valid by the State in which it is registered.

Article 32

Licenses of personnel

a) The pilot of every aircraft and the other members of the operating crew of every aircraft engaged in international navigation shall be provided with certificates of competency and licenses issued or rendered valid by the State in which the aircraft is registered.
b) Each contracting State reserves the right to refuse to recognize, for the purpose of flight above its own territory, certificates of competency and licenses granted to any of its nationals by another contracting State.

Article 33

Recognition of certificates and licenses

Certificates of airworthiness and certificates of competency and licenses issued or rendered valid by the contracting State in which the aircraft is registered, shall be recognized as valid by the other contracting States, provided that the requirements under which such certificates or licenses were issued or rendered valid are equal to or above the minimum standards which may be established from time to time pursuant to this Convention.

Article 34

Journey log books

There shall be maintained in respect of every aircraft engaged in international navigation a journey log book in which shall be entered particulars of the aircraft, its crew and of each journey, in such form as may be prescribed from time to time pursuant to this Convention.

Chapter VI: International standards and recommended practices

Article 37

Adoption of international standards and procedures

Each contracting State undertakes to collaborate in securing the highest practicable degree of uniformity in regulations, standards, procedures, and organization in relation to aircraft, personnel, airways and auxiliary services in all matters in which such uniformity will facilitate and improve air navigation.

To this end the International Civil Aviation Organization shall adopt and amend from time to time, as may be necessary, international standards and recommended practices and procedures dealing with:

(a) Communications systems and air navigation aids, including ground marking;
(b) Characteristics of airports and landing areas;
(c) Rules of the air and air traffic control practices;
(d) Licensing of operating and mechanical personnel;
(e) Airworthiness of aircraft;
(f) Registration and identification of aircraft;
(g) Collection and exchange of meteorological information;
(h) Log books;
(i) Aeronautical maps and charts;
(j) Customs and immigration procedures;
(k) Aircraft in distress and investigation of accidents;

and such other matters concerned with the safety, regularity, and efficiency of air navigation as may from time to time appear appropriate.

Article 38

Departure from international standards and procedures

Any State which finds it impracticable to comply in all respects with any such international standards or procedure, or to bring its own regulations or practices into full accord with any international standard or procedure after amendment of the latter, or which deems it necessary to adopt regulations or practices differing in any particular respect from those established by an international standard, shall give immediate notification to the International Civil Aviation Organization of the differences between its own practice and that established by the international standard. In the case of amendments to international standards, any State which does not make the appropriate amendments to its own regulations or practices shall give notice to the Council within 60 days of the adoption of the amendment to the international standard, or indicate the action which it proposes to take. In any such case, the Council shall make immediate notification to all other States of the difference which exists between one or more features of an international standard and the corresponding national practice of that State.

Article 39

Endorsement of certificates and licenses

(a) Any aircraft or part thereof with respect to which there exists an international standard of airworthiness or performance, and which failed in any respect to satisfy that standard at the time of its certification, shall have endorsed on or attached to its airworthiness certificate a complete enumeration of the details in respect of which it so failed.

(b) Any person holding a license who does not satisfy in full the conditions laid down in the international standard relating to the class of license or certificate which he holds shall have endorsed on or attached to his license a complete enumeration of the particulars in which he does not satisfy such conditions.

Article 40

Validity of endorsed certificates and licenses

No aircraft or personnel having certificates or licenses so endorsed shall participate in international navigation, except with the permission of the State or States whose territory is entered. The registration or use of any such aircraft, or of any certificated aircraft part, in any State other than that in which it was originally certificated shall be at the discretion of the State into which the aircraft or part is imported.

Article 41

Recognition of exiting standards of airworthiness

The provisions of this Chapter shall not apply to aircraft and aircraft equipment of types of which the prototype is submitted to the appropriate national authorities for certification prior to a date three years after the date of adoption of an international standard of airworthiness for such equipment.

Article 42

Recognition of exiting standards of competency of personnel

The provisions of this Chapter shall not apply to personnel whose licenses are originally issued prior to a date one year after initial adoption of an international standard of qualification for such personnel; but they shall in any case apply to all personnel whose licenses remain valid five years after the date of adoption of such standard.

Excerpts from Part II: The international civil aviation organization

Chapter VII: The organization

Article 44

Objectives

The aims and objectives of the Organization are to develop the principles and techniques of international air navigation and to foster the planning and development of international air transport so as to:

(a) Insure the safe and orderly growth of international civil aviation throughout the world;
(b) Encourage the arts of aircraft design and operation for peaceful purposes;
(c) Encourage the development of airways, airports, and air navigation facilities for international civil aviation;
(d) Meet the needs of the peoples of the world for safe, regular, efficient and economical air transport;
(e) Prevent economic-waste caused by unreasonable competition;
(f) Insure that the rights of contracting States are fully respected and that every contracting State has a fair opportunity to operate international airlines;
(g) Avoid discrimination between contracting States;
(h) Promote safety of flight in international air navigation;
(i) Promote generally the development of all aspects of international civil aeronautics.

Excerpts from Part III: International air transport

Chapter XIV: Information and reports

Article 67

File reports with Council

Each contracting State undertakes that its international airlines shall, in accordance with requirements laid down by the Council, file with the Council traffic reports, cost statistics and financial statements showing among other things all receipts and the sources thereof.

Chapter XV: Airports and other air navigation facilities

Article 69

Improvement of air navigation facilities

If the Council is of the opinion that the airports or other air navigation facilities, including radio and meteorological services, of a contracting State are not reasonably adequate for the safe, regular, efficient, and economical operation of international air services, present or contemplated, the Council shall consult with the State directly concerned, and other States affected, with a view to finding means by which the situation may be remedied, and may make recommendations for that purpose. No contracting State shall be guilty of an infraction of this Convention if it fails to carry out these recommendations.

Article 71

Provision and maintenance of facilities by Council

If a contracting State so requests, the Council may agree to provide, man, maintain, and administer any or all of the airports and other air navigation facilities, including radio and meteorological services, required in its territory for the safe, regular, efficient and economical operation of the international air services of the other contracting States, and may specify just and reasonable charges for the use of the facilities provided.

Relevant excerpts from Annex 1 to Chicago convention – personnel licensing

Chapter 1: Definitions and general rules concerning licences (excerpts)

1.2 General rules concerning licences

Note 2. – International Standards and Recommended Practices are established for licensing the following personnel:

a) Flight crew

- private pilot – aeroplane, airship, helicopter or powered-lift;
- commercial pilot – aeroplane, airship, helicopter or powered-lift;

- multi-crew pilot – aeroplane;
- airline transport pilot – aeroplane, helicopter or powered-lift;
- glider pilot;
- free balloon pilot;
- flight navigator;
- flight engineer.

b) Other personnel

- aircraft maintenance (technician/engineer/mechanic);
- air traffic controller;
- flight operations officer/flight dispatcher;
- aeronautical station operator.

1.2.1 Authority to act as a flight crew member
A person shall not act as a flight crew member of an aircraft unless a valid licence is held showing compliance with the specifications of this Annex and appropriate to the duties to be performed by that person. The licence shall have been issued by the State of Registry of that aircraft or by any other Contracting State and rendered valid by the State of Registry of that aircraft.

1.2.2.3 **Recommendation** – *A pilot licence issued by a Contracting State should be rendered valid by other Contracting States for use in private flights.*

1.2.4 Medical fitness

1.2.4.1 An applicant for a licence shall, when applicable, hold a Medical Assessment issued in accordance with the provisions of Chapter 6.

1.2.5 Validity of licences

1.2.5.1 A Contracting State, having issued a licence, shall ensure that the privileges granted by that licence, or by related ratings, are not exercised unless the holder maintains competency and meets the requirements for recent experience established by that State.

1.2.5.1.1 **Recommendation** – A Contracting State should establish maintenance of competency and recent experience requirements for pilot licences and ratings based on a systematic approach to accident prevention and should include a risk assessment process and analysis of current operations, including accident and incident data appropriate to that State.

Chapter 2: Licences and Ratings for Pilots (Excerpts)

2.1 General rules concerning pilot licences and ratings

2.1.1 General licensing specifications

2.1.1.1 A person shall not act either as pilot-in-command or as co-pilot of an aircraft in any of the following categories unless that person is the holder of a pilot licence issued in accordance with the provisions of this Chapter:

- aeroplane
- airship of a volume of more than 4600 cubic metres
- free balloon
- glider
- helicopter
- powered-lift.

2.4 Commercial pilot licence

2.4.1 General requirements for the issue of the licence appropriate to the aeroplane, airship, helicopter and powered-lift categories.

2.4.1.1 *Age*
The applicant shall be not less than 18 years of age.

2.4.1.2 *Knowledge*
The applicant shall have demonstrated a level of knowledge appropriate to the privileges granted to the holder of a commercial pilot licence and appropriate to the category of aircraft intended to be included in the licence, in at least the following subjects:

Air law

a) rules and regulations relevant to the holder of a commercial pilot licence; rules of the air; appropriate air traffic services practices and procedures;

Aircraft general knowledge for aeroplanes, airships, helicopters and powered-lifts

b) principles of operation and functioning of powerplants, systems and instruments;

c) operating limitations of the relevant category of aircraft and powerplants; relevant operational information from the flight manual or other appropriate document;

d) use and serviceability checks of equipment and systems of appropriate aircraft;

e) maintenance procedures for airframes, systems and powerplants of appropriate aircraft;

Flight performance, planning and loading

h) effects of loading and mass distribution on aircraft handling, flight characteristics and performance; mass and balance calculations;

i) use and practical application of take-off, landing and other performance data;

j) pre-flight and en-route flight planning appropriate to commercial operations under VFR; preparation and filing of air traffic services flight plans; appropriate air traffic services procedures; altimeter setting procedures;

k) in the case of airships, helicopters and powered-lifts, effects of external loading on handling;

Human performance

l) human performance including principles of threat and error management;

Meteorology

m) interpretation and application of aeronautical meteorological reports, charts and forecasts; use of, and procedures for obtaining, meteorological information, pre-flight and in-flight; altimetry;

n) aeronautical meteorology; climatology of relevant areas in respect of the elements having an effect upon aviation; the movement of pressure systems, the structure of fronts, and the origin and characteristics of significant weather phenomena which affect take-off, en-route and landing conditions;

o) causes, recognition and effects of icing; frontal zone penetration procedures; hazardous weather avoidance;

Navigation

p) air navigation, including the use of aeronautical charts, instruments and navigation aids; an understanding of the principles and characteristics of appropriate navigation systems; operation of airborne equipment;

Operational procedures

r) application of threat and error management to operational performance;

u) appropriate precautionary and emergency procedures;
v) operational procedures for carriage of freight; potential hazards associated with dangerous goods;

Principles of flight

y) principles of flight;

Radiotelephony

z) communication procedures and phraseology as applied to VFR operations; action to be taken in case of communication failure.

2.4.1.3 *Skill*

The applicant shall have demonstrated the ability to perform as pilot-in-command of an aircraft within the appropriate category of aircraft, the procedures and manoeuvres described in 2.4.3.2 or 2.4.4.2 or 2.4.5.2 or 2.4.6.2 with a degree of competency appropriate to the privileges granted to the holder of a commercial pilot licence, and to:

a) recognize and manage threats and errors;
b) operate the aircraft within its limitations;
c) complete all manoeuvres with smoothness and accuracy;
d) exercise good judgement and airmanship;
e) apply aeronautical knowledge; and
f) maintain control of the aircraft at all times in a manner such that the successful outcome of a procedure or manoeuvre is assured.

2.4.1.4 *Medical fitness*

The applicant shall hold a current Class 1 Medical Assessment.

2.4.3.2 *Flight instruction*

The applicant shall have received dual instruction in aeroplanes appropriate to the class and/or type rating, sought from an authorized flight instructor.

Relevant excerpts from Annex 6 to the Chicago convention – operation of aircraft

Chapter 3: General (excerpts)

3.1 Compliance with laws, regulations and procedures

3.1.1 An operator shall ensure that all employees when abroad know that they must comply with the laws, regulations and procedures of those States in which operations are conducted.

3.1.2 An operator shall ensure that all pilots are familiar with the laws, regulations and procedures, pertinent to the performance of their duties, prescribed for the areas to be traversed, the aerodromes to be used and the air navigation facilities relating thereto. The operator shall ensure that other members of the flight crew are familiar with such of these laws, regulations and procedures as are pertinent to the performance of their respective duties in the operation of the aeroplane.

3.2 Safety management

3.2.1 States shall establish a safety programme in order to achieve an acceptable level of safety in the operation of aircraft.

3.2.2 The acceptable level of safety to be achieved shall be established by the State(s) concerned.

3.2.3 **Recommendation** – States should require, as part of their safety programme, that an operator implement a safety management system acceptable to the State of the Operator that, as a minimum:

a) identifies safety hazards;
b) ensures that remedial action necessary to maintain an acceptable level of safety is implemented;
c) provides for continuous monitoring and regular assessment of the safety level achieved; and
d) aims to make continuous improvement to the overall level of safety.

3.2.4 From 1 January 2009, States shall require, as part of their safety programme, that an operator implement a safety management system acceptable to the State of the Operator that, as a minimum:

a. identifies safety hazards;
b. ensures that remedial action necessary to maintain an acceptable level of safety is implemented;
c. provides for continuous monitoring and regular assessment of the safety level achieved and
d. aims to make continuous improvement to the overall level of safety.

3.2.5 A safety management system shall clearly define lines of safety accountability throughout the operator's organization, including a direct accountability for safety on the part of senior management.

3.2.6 **Recommendation** – An operator of an aeroplane of a certificated take-off mass in excess of 20,000 kg should establish and maintain a flight data analysis programme as part of its safety management system.

3.2.7 An operator of an aeroplane of a maximum certificated take-off mass in excess of 27,000 kg shall establish and maintain a flight data analysis programme as part of its safety management system.

3.2.8 A flight data analysis programme shall be non-punitive and contain adequate safeguards to protect the source(s) of the data.

3.2.9 An operator shall establish a flight safety documents system, for the use and guidance of operational personnel, as part of its safety management system.

Chapter 4: Flight Operations (excerpts)

4.1 Operating facilities

4.1.1 An operator shall ensure that a flight will not be commenced unless it has been ascertained by every reasonable means available that the ground and/or water facilities available and directly required on such flight, for the safe operation of the aeroplane and the protection of the passengers, are adequate for the type of operation under which the flight is to be conducted and are adequately operated for this purpose.

4.1.2 An operator shall ensure that any inadequacy of facilities observed in the course of operations is reported to the authority responsible for them, without undue delay.

4.1.3 Subject to their published conditions of use, aerodromes and their facilities shall be kept continuously available for flight operations during their published hours of operations, irrespective of weather conditions.

4.2 Operational certification and supervision

4.2.1 The air operator certificate

4.2.1.1 An operator shall not engage in commercial air transport operations unless in possession of a valid air operator certificate issued by the State of the Operator.

4.2.1.2 The air operator certificate shall authorize the operator to conduct commercial air transport operations in accordance with specified authorizations, conditions and limitations.

4.2.1.3 Contracting States shall recognize as valid an air operator certificate issued by another Contracting State, provided that the requirements under which the certificate was issued are at least equal to the applicable Standards specified in this Annex.

4.2.1.4 The issue of an air operator certificate by the State of the Operator shall be dependent upon the operator demonstrating an adequate organization, method of control and supervision of flight operations, training programme as well as ground handling and maintenance arrangements consistent with the nature and extent of the operations specified.

4.2.1.5 The continued validity of an air operator certificate shall depend upon the operator maintaining the requirements of 4.2.1.4 under the supervision of the State of the Operator.

4.2.1.6 The air operator certificate shall contain at least the following:

a) operator's identification (name, location);
b) date of issue and period of validity;
c) description of the types of operations authorized;
d) the type(s) of aircraft authorized for use; and
e) authorized areas of operation or routes.

4.2.1.7 The State of the Operator shall establish a system for both the certification and the continued surveillance of the operator in accordance with Appendix 5 to ensure that the required standards of operations established in 4.2 are maintained.

Relevant excerpts from Annex 8 to the Chicago convention: airworthiness of aircraft

Chapter 1: Type Certification (excerpts)

1.1 Applicability

The Standards of this chapter shall be applicable to all aircraft of types for which the application for certification was submitted to a Contracting State on or after 13 June 1960, except that the provisions of 1.4 of this part shall only be applicable to an aircraft type for which an application for a Type Certificate is submitted to the State of Design on or after 2 March 2004.

1.2 Design aspects of the appropriate airworthiness requirements

1.2.1 The design aspects of the appropriate airworthiness requirements, used by a Contracting State for type certification in respect of a class of aircraft or for any change to such type certification, shall be such that compliance with them will ensure compliance with the Standards of Part II of this Annex and, where applicable, with the Standards of Parts IIIA, IIIB and IV of this Annex.

1.2.2 The design shall not have any features or characteristics that render it unsafe under the anticipated operating conditions.

1.2.3 Where the design features of a particular aircraft render any of the design aspects of the appropriate airworthiness requirements or the Standards in Parts IIIA, IIIB and IV inappropriate, the Contracting State shall apply appropriate requirements that will give at least an equivalent level of safety.

1.2.4 Where the design features of a particular aircraft render any of the design aspects of the appropriate airworthiness requirements or the Standards in Parts IIIA, IIIB and IV inadequate, additional requirements that are considered by the Contracting State to give at least an equivalent level of safety shall be applied.

1.3 Proof of compliance with the appropriate airworthiness requirements

1.3.1 There shall be an approved design consisting of such drawings, specifications, reports and documentary evidence as are necessary to define the design of the aircraft and to show compliance with the design aspects of the appropriate airworthiness requirements.

1.3.2 The aircraft shall be subjected to such inspections and ground and flight tests as are deemed necessary by the State to show compliance with the design aspects of the appropriate airworthiness requirements.

1.3.3 In addition to determining compliance with the design aspects of the appropriate airworthiness requirements for an aircraft, Contracting States shall take whatever other steps they deem necessary to ensure that the design approval is withheld if the aircraft is known or suspected to have dangerous features not specifically guarded against by those requirements.

1.3.4 A Contracting State issuing an approval for the design of a modification, of a repair or of a replacement part shall do so on the basis of satisfactory evidence that the aircraft is in compliance with the airworthiness requirements used for the issuance of the Type Certificate, its amendments or later requirements when determined by the State.

1.4 Type Certificate

1.4.1 The State of Design, upon receipt of satisfactory evidence that the aircraft type is in compliance with the design aspects of the appropriate airworthiness requirements, shall issue a Type Certificate to define the design and to signify approval of the design of the aircraft type.

1.4.2 When a Contracting State, other than the State of Design, issues a Type Certificate for an aircraft type, it shall do so on the basis of satisfactory evidence that the aircraft type is in compliance with the design aspects of the appropriate airworthiness requirements.

Chapter 2: Production (excerpts)

2.1 Applicability

The Standards of this chapter are applicable to all aircraft.

2.2 Production

2.2.1 Aircraft production
The State of Manufacture shall ensure that each aircraft, including parts manufactured by sub-contractors, conforms to the approved design.

2.2.2 Parts production
The Contracting State taking responsibility for the production of parts manufactured under the design approval referred to in 1.3.4 of Part II shall ensure that the parts conform to the approved design.

2.2.3 Production control

When approving production of aircraft or aircraft parts, a Contracting State shall ensure that it is performed in a controlled manner including the use of a quality system so that construction and assembly are satisfactory.

2.2.4 Traceability

Records shall be maintained such that the identification of the aircraft and of the parts with their approved design and production can be established.

Chapter 3: Certificate of Airworthiness (excerpts)

3.1 Applicability

The Standards of this chapter are applicable in respect of all aircraft, except 3.3 and 3.4 which are not applicable in respect of all aircraft that are of a type of which the prototype was submitted to appropriate national authorities for certification before 13 June 1960.

3.2 Issuance and continued validity of a Certificate of Airworthiness

3.2.1 A Certificate of Airworthiness shall be issued by a Contracting State on the basis of satisfactory evidence that the aircraft complies with the design aspects of the appropriate airworthiness requirements.

3.2.2 A Contracting State shall not issue or render valid a Certificate of Airworthiness for which it intends to claim recognition pursuant to Article 33 of the Convention on International Civil Aviation unless it has satisfactory evidence that the aircraft complies with the applicable Standards of this Annex through compliance with appropriate airworthiness requirements.

3.2.3 A Certificate of Airworthiness shall be renewed or shall remain valid, subject to the laws of the State of Registry, provided that the State of Registry shall require that the continuing airworthiness of the aircraft shall be determined by a periodical inspection at appropriate intervals having regard to lapse of time and type of service or, alternatively, by means of a system of inspection, approved by the State, that will produce at least an equivalent result.

3.4 Aircraft limitations and information

Each aircraft shall be provided with a flight manual, placards or other documents stating the approved limitations within which the aircraft is considered airworthy as defined by the appropriate airworthiness requirements and additional instructions and information necessary for the safe operation of the aircraft.

Part III: Large aeroplanes

Part IIIA: Aeroplanes over 5700 kg for which application for certification was submitted on or after 13 June 1960 but before 2 March 2004

Chapter 1: General (excerpts)

1.4 Unsafe features and characteristics

The aeroplane shall not possess any feature or characteristic that renders it unsafe under the anticipated operating conditions.

1.5 Proof of compliance

1.5.1 Compliance with the appropriate airworthiness requirements shall be based on evidence either from tests, calculations, or calculations based on tests, provided that in each case the accuracy achieved will ensure a level of airworthiness equal to that which would be achieved were direct tests conducted.

1.5.2 The tests of 1.5.1 shall be such as to provide reasonable assurance that the aeroplane, its components and equipment are reliable and function correctly under the anticipated operating conditions.

Chapter 2: Flight (excerpts)

2.1 General

2.1.1 Compliance with the Standards prescribed in Chapter 2 shall be established by flight or other tests conducted upon an aeroplane or aeroplanes of the type for which a Certificate of Airworthiness is sought, or by calculations based on such tests, provided that the results obtained by calculations are equal in accuracy to, or conservatively represent, the results of direct testing.

2.2 Performance

2.2.1 General

2.2.1.1 Sufficient data on the performance of the aeroplane shall be determined and scheduled in the flight manual to provide operators with the necessary information for the purpose of determining the total mass of the aeroplane on the basis of the values, peculiar to the proposed flight, of the relevant operational parameters, in order that the flight may be made with reasonable assurance that a safe minimum performance for that flight will be achieved.

2.2.1.2 The performance scheduled for the aeroplane shall take into consideration human performance and in particular shall not require exceptional skill or alertness on the part of the flight crew.

2.3 Flying qualities

The aeroplane shall comply with the Standards of 2.3 at all altitudes up to the maximum anticipated altitude relevant to the particular requirement in all temperature conditions relevant to the altitude in question and for which the aeroplane is approved.

2.3.1 Controllability

The aeroplane shall be controllable and manoeuvrable under all anticipated operating conditions, and it shall be possible to make smooth transitions from one flight condition to another (e.g. turns, sideslips, changes of engine power, changes of aeroplane configurations) without requiring exceptional skill, alertness or strength on the part of the pilot even in the event of failure of any power-unit. A technique for safely controlling the aeroplane shall be established for all stages of flight and aeroplane configurations for which performance is scheduled.

Chapter 3: Structures (excerpts)

3.1 General

The Standards of Chapter 3 apply to the aeroplane structure consisting of all portions of the aeroplane, the failure of which would seriously endanger the aeroplane.

Chapter 4: Design and Construction (excerpts)

4.1 General

Details of design and construction shall be such as to give reasonable assurance that all aeroplane parts will function effectively and reliably in the anticipated operating conditions. They shall be based upon practices that experience has proven to be satisfactory or that are substantiated by special tests or by other appropriate investigations or both. They shall observe Human Factors principles.

4.1.1 Substantiating tests

The functioning of all moving parts essential to the safe operation of the aeroplane shall be demonstrated by suitable tests in order to ensure that they will function correctly under all operating conditions for such parts.

4.1.2 Materials

All materials used in parts of the aeroplane essential for its safe operation shall conform to approved specifications. The approved specifications shall be such that materials accepted as complying with the specifications will have the essential properties assumed in the design.

4.1.3 Fabrication methods

The methods of fabrication and assembly shall be such as to produce a consistently sound structure which shall be reliable with respect to maintenance of strength in service.

4.1.4 Protection

The structure shall be protected against deterioration or loss of strength in service due to weathering, corrosion, abrasion or other causes, which could pass unnoticed, taking into account the maintenance the aeroplane will receive.

4.1.5 Inspection provisions

Adequate provision shall be made to permit any necessary examination, replacement or reconditioning of parts of the aeroplane that require such attention, either periodically or after unusually severe operations.

Chapter 8: Instruments and Equipment (excerpts)

8.1 Required instruments and equipment

The aeroplane shall be provided with approved instruments and equipment necessary for the safe operation of the aeroplane in the anticipated operating conditions. These shall include the instruments and equipment necessary to enable the crew to operate the aeroplane within its operating limitations.

8.3 Safety and survival equipment

Prescribed safety and survival equipment that the crew or passengers are expected to use or operate at the time of an emergency shall be reliable, readily accessible and easily identified, and its method of operation shall be plainly marked.

Chapter 10: Continuing Airworthiness – Maintenance Information (excerpts)

10.1 General

Information for use in developing procedures for maintaining the aeroplane in an airworthy condition shall be made available. The information shall include that described in 10.2, 10.3 and 10.4.

10.2 Maintenance information

Maintenance information shall include a description of the aeroplane and recommended methods for the accomplishment of maintenance tasks. Such information shall include guidance on defect diagnosis.

10.3 Maintenance programme information

Maintenance programme information shall include the maintenance tasks and the recommended intervals at which these tasks are to be performed.

10.4 Maintenance information resulting from the type design approval

Maintenance tasks and frequencies that have been specified as mandatory by the State of Design in approval of the type design shall be identified as such.

Appendix C

Model code of conduct for space-faring nations[2]

Released by the Stimson Centre on October 24, 2007

Central Objective of this Code of Conduct

To preserve and advance the peaceful exploration and use of outer space.

Preamble

We the undersigned;

Recognizing the common interest of all humankind in achieving progress in the exploration and use of outer space for peaceful purposes;

Reaffirming the crucial importance of outer space for global economic progress, commercial advancement, scientific research, sustainable development, as well as national, regional and international security;

Desiring to prevent conflict in outer space;

Reaffirming our commitment to the United Nations Charter;

Taking into consideration the salience of Article 2(4) of the Charter, which obliges all members to refrain in their international relations from the threat or use of force against the territorial integrity or political independence of any state, or in any other manner inconsistent with the purposes of the United Nations;

Taking special account of Article 42 of the Charter, under which the United Nations Security Council may mandate action by air, sea or land forces as may be necessary to maintain or restore international peace and security;

Recognizing the inherent right of self-defense of all states under Article 51 of the Charter;

Reinforcing the principles of the Outer Space Treaty of 1967, including:

- the exploration and use of outer space, including the moon and other celestial bodies, shall be carried out for the benefit and in the interests of all countries,

- outer space, including the moon and other celestial bodies, shall be free for exploration and use by all States without discrimination of any kind, on a basis of equality and in accordance with international law;
- outer space, including the moon and other celestial bodies, is not subject to national appropriation by claim of sovereignty, by means of use or occupation, or by any other means, in the exploration and use of outer space, States Parties to the Treaty shall be guided by the principle of co-operation and mutual assistance and shall conduct all their activities in outer space with due regard to the corresponding interests of all other States Parties to the Treaty;
- State Parties to the Treaty undertake not to place in orbit around the Earth any objects carrying weapons of mass destruction;
- the moon and other celestial bodies shall be used by all States Parties to the Treaty exclusively for peaceful purposes.

Recalling the importance of space assets for non-proliferation, disarmament and arms control treaties, conventions and regimes;

Recognizing that harmful actions against space objects would have injurious consequences for international peace, security and stability;

Encouraging signature, ratification, accession, and adherence to all legal instruments governing outer space, including:

- 1967 Outer Space Treaty
- 1968 Rescue Agreement
- 1972 Liability Convention
- 1976 Registration Convention
- 1984 Moon Agreement

Recognizing the value of mechanisms currently in place related to outer space, including the 1994 Constitution of International Telecommunications Union; the 1963 Partial Test Ban Treaty; the 1988 Intermediate-Range Nuclear Forces Treaty; the 1994 Strategic Arms Reduction Treaty; and the 2003 Treaty on Strategic Offensive Reductions;

Recognizing the dangers posed by space debris for safe space operations and recognizing the importance of the 2007 Space Debris Mitigation Guidelines of the Scientific and Technical Sub-committee of the Committee on the Peaceful Uses of Outer Space;

Recognizing the importance of a space traffic management system to assist in the safe and orderly operation of outer space activities;

Believing that universal adherence to this Code of Conduct does not in any way diminish the need for additional international legal instruments that preserve, advance and guarantee the exploration and use of outer space for peaceful purposes; Declare the following rights and responsibilities:

Rights of Space-Faring States:

1. The right of access to space for exploration or other peaceful purposes.
2. The right of safe and interference-free space operations, including military support functions.
3. The right of self-defense as enumerated in the Charter of the United Nations.
4. The right to be informed on matters pertaining to the objectives and purposes of this Code of Conduct.
5. The right of consultation on matters of concern and the proper implementation of this Code of Conduct.

Responsibilities of Space-Faring States:

1. The responsibility to respect the rights of other space-faring states and legitimate stakeholders.
2. The responsibility to regulate stakeholders that operate within their territory or that use their space launch services in conformity with the objectives and purposes of this Code of Conduct.
3. Each state has the responsibility to regulate the behavior of its nationals in conformity with the objectives and purposes of this Code of Conduct, wherever those actions occur.
4. The responsibility to develop and abide by rules of safe space operation and traffic management.
5. The responsibility to share information related to safe space operations and traffic management and to enhance cooperation on space situational awareness.
6. The responsibility to mitigate and minimize space debris in accordance with the best practices established by the international community in such agreements as the Inter-Agency Debris Coordination Committee guidelines and guidelines of the Scientific and Technical Sub-committee of the United Nations Committee on the Peaceful Uses of Outer Space.
7. The responsibility to refrain from harmful interference against space objects.
8. The responsibility to consult with other space-faring states regarding activities of concern in space and to enhance cooperation to advance the objectives and purposes of this Code of Conduct.

9. The responsibility to establish consultative procedures to address and resolve questions relating to compliance with this Code of Conduct, and to agree upon such additional measures as may be necessary to improve the viability and effectiveness of this Code of Conduct.

The Model Code of Conduct was completed by experts from NGOs in Canada, France, Japan, Russia and the United States in October 2007. The group included Setsuko Aoki of Keio University, Alexei Arbatov of the Carnegie Moscow Center, Vladimir Dvorkin of the Center for Policy Studies in Russia, Trevor Findlay of the Canadian Centre for Treaty Compliance, Katsuhisa Furukawa of the Japan Science and Technology Agency, Scott Lofquist-Morgan of the Canadian Centre for Treaty Compliance, Laurence Nardon of the French Institute of International Relations, and Sergei Oznobistchev of the Institute of Strategic Studies and Analysis. NGO participants worked on this project in a personal capacity. Their support for the model Code of Conduct therefore does not reflect endorsements by their institutions or governments.

[2] Available at: http://www.stimson.org/books-reports/model-code-of-conduct-for-space-faring-nations/ (last accessed: 03 January 2011).

About the editors

Prof. Ram S. Jakhu has over 25 years of experience in space-related fields. He holds a position of Associate Professor at the Institute of Air and Space Law, Faculty of Law, of McGill University in Montreal, Canada, where he teaches several courses covering numerous subjects including, *inter alia*, public international law, international and national space law and policy, international trade, export controls, space applications, space commercialization, and telecommunications. From January 1995 to December 1998, Dr. Jakhu served the International Space University, Strasbourg, France, in a full-time capacity holding various titles, including Professor and first Director of the Master of Space Studies Programme. He has authored more than sixty articles in several reputed journals, and edited two books on "Space Safety Regulations and Standards" and "National Regulation of Space Activities." He has presented numerous papers and expert legal opinions at various conferences around the world and participated in several space-related studies. Prof. Jakhu is a "Fellow" as well as the Chairman of the Legal and Regulatory Committee of the International Association for the Advancement of Space Safety. He is a Member of the Board of Directors of the International Institute of Space law of the International Astronautical Federation (Paris). In 2007, he received a "Distinguished Service Award" from the International Institute of Space Law for his significant contribution to the development of space law. He holds a Doctor of Civil Law (Dean's Honours List) degree in Space Law from McGill University and a Master of Law (LL.M.) degree in the field of Air and Space Law from McGill University. In addition, he has earned LL.M. (in Public and Private International Law), LL.B. (in Laws of India) and Bachelor of Arts (in Economics and Political Science) degrees from Panjab University, Chandigarh, India.

Tommaso Sgobba holds an M.S. in Aeronautical Engineering from the Polytechnic of Turin (I), where he has been also professor of space system safety (1999–2001). T. Sgobba has over 33 years of experience in the aerospace industry. He is currently a staff member of the European Space Agency in charge of flight

safety for manned systems, spacecraft re-entries safety, space debris, use of nuclear power sources, and planetary protection. T. Sgobba joined the European Space Agency in 1989, after thirteen years in the aeronautical industry. Initially, he supported the development of the Ariane 5 launcher, of Earth observation and meteorological satellites, and the early Hermes space plane phase. Later, he became product assurance and safety manager for all European manned missions on Shuttle, MIR station, and for the European research facilities for the International Space Station. During his long and close cooperation with the NASA Shuttle/ISS Payload Safety Review Panel, T. Sgobba developed at ESA the safety technical and organizational capabilities that eventually led to the establishment of the first ESA formal safety review panel and first International Partner ISS Payload Safety Review Panel in 2002. He was also instrumental in setting up the ESA ATV Re-entry Safety Panel and to organize the first scientific observation campaign of a destructively re-entering spacecraft (ATV- Jules Verne). T. Sgobba has published several papers on space safety, and has co-edited the text book "Safety Design for Space Systems", published in 2009 by Elsevier. Mr. Sgobba received NASA recognition for his outstanding contribution to the International Space Station in 2004, and the prestigious NASA Space Flight Awareness (SFA) Award in 2007. T. Sgobba is President and co-founder of the IAASS (International Association for the Advancement of Space Safety). He is also vice-president and co-founder of the U.S. based International Space Safety Foundation (ISSF).

Dr. Paul Stephen Dempsey is Tomlinson Professor of Global Governance in Air and Space Law and Director of the Institute of Air and Space Law at McGill University, in Montreal, Canada. From 1979 to 2002, he held the endowed chair as Professor of Transportation Law and Director of the Transportation Law Programme at the University of Denver. He was also Director of the National Center for Intermodal Transportation. From 1975 to 1979, he served as an attorney with the Civil Aeronautics Board and the Interstate

Commerce Commission in Washington, D.C., and in 1981–81 he was Legal Advisor to the Chairman of the I.C.C. Professor Dempsey has written nearly 100 law review and professional journal articles, scores of newspaper and news magazine editorials, and twenty books. Dr. Dempsey holds the following degrees: Bachelor of Arts (1972), Juris Doctor (1975), University of Georgia; Master of Laws (1978), George Washington University; Doctor of Civil Laws (1986), McGill University. He is admitted to practice law in Colorado, Georgia and the District of Columbia. Professor Dempsey was a Fulbright Scholar, was awarded the Transportation Lawyers Association Distinguished Service Award, and was designated the University of Denver's Outstanding Scholar. He was the first individual designated the University of Denver's Hughes Research Professor and DePaul University's Distinguished Visiting Professor of Law. The Colorado transportation community named him "Educator of the Year", and inducted him into the Colorado Aerospace Hall of Fame. From 1979 to 2002, he was faculty editor of the Transportation Law Journal. He also served on the Editorial Boards of the Denver Business Journal, and The Aviation Quarterly (Lloyds, London).

International Association for the Advancement of Space Safety

The International Association for the Advancement of Space Safety (IAASS), legally established 16 April 2004 in the Netherlands, is a non-profit organization dedicated to furthering international cooperation and scientific advancement in the field of space systems safety. IAASS membership is open to anyone having a professional interest in space safety. For more information, visit: http://www.iaass.org/.

Institute of Air and Space Law, McGill University

In 1951, McGill University established the Institute of Air and Space Law (IASL) to provide graduate legal education to students from around the world. In the ensuing half century, IASL has educated more than 900 students, who today occupy senior positions in some of the most well known law firms, corporations, governmental and intergovernmental institutions in more than 120 countries around the world. In 1996, on the occasion of its 45th anniversary, the Institute received the prestigious Edward Warner Award, the highest distinction in the field of civil aviation awarded by the Council of ICAO. The award was granted "in recognition of the Institute's significant contribution to the development of international air law". For more information, visit: http://www.mcgill.ca/iasl/.

MIX
Papier aus verantwortungsvollen Quellen
Paper from responsible sources
FSC® C105338

If you have any concerns about our products,
you can contact us on
ProductSafety@springernature.com

In case Publisher is established outside the EU,
the EU authorized representative is:
Springer Nature Customer Service Center GmbH
Europaplatz 3, 69115 Heidelberg, Germany

Printed by Libri Plureos GmbH
in Hamburg, Germany